Carl Robert Osten-Sacken

Western Diptera

Descriptions of New Genera and Species of Diptera from the Region West of the

Mississippi and Especially from California

Carl Robert Osten-Sacken

Western Diptera
*Descriptions of New Genera and Species of Diptera from the Region West of the Mississippi
and Especially from California*

ISBN/EAN: 9783337805470

Printed in Europe, USA, Canada, Australia, Japan

Cover: Foto ©berggeist007 / pixelio.de

More available books at **www.hansebooks.com**

TABLE OF CONTENTS

ERRATA.

Page 228, line 19 from top, remove the word *Anisotamia* to the beginning of the preceding line.

Page 230, line 13 from top, for *posterior* read *submarginal*.

Page 267, line 16 from bottom, for *fucata* read *amphitea*.

III.—WESTERN DIPTERA. DESCRIPTIONS OF NEW GENERA AND SPECIES OF DIPTERA FROM THE REGION WEST OF THE MISSISSIPPI AND ESPECIALLY FROM CALIFORNIA.

By C. R. Osten Sacken.

PREFACE.

The *Diptera* of the Pacific coast are at present almost unknown. A few species picked up during the visit of the Swedish frigate Eugenia, probably in the environs of San Francisco, and described by Mr. Thomson; some four dozen species, published by Mr. Loew in his "Centuriæ"; a few other species, by Dr. Gerstaecker; and two *Limnobiæ*, by me, constitute about all we know of Californian *Diptera*. Even Chili is, in this respect, much better explored, with the 556 species contained in Dr. Philippi's publication.

In the present publication, I give a survey of the collection of *Diptera* which I formed during my recent western journey, and describe the most remarkable forms. The majority of the species described belong to California, where I collected the most; the fauna of Colorado and of the vast intermediate region will come in the second line only, the materials being less abundant. However, the more I proceed with my study, the more I am impressed with the fact that the western fauna is essentially *one*, and that many of the characteristic forms of California sooner or later will turn up in Colorado.

The times and places of my collecting in California are as follows: During the winter months (January to March, 1875), I collected a little in Southern California; my most active collecting, however, was confined to the months of April and May, 1876, in Marin and Sonoma Counties; a few days in Yosemite Valley in June; and a couple of weeks in the Sierra Nevada in July, especially about Webber Lake, Sierra County. What I brought together is therefore but a small fragment of the fauna, collected during a very limited season. And, yet, even this fragment yields some very interesting facts concerning the geographical distribution of insects, discloses unexpected analogies and coincidences between the fauna of California and those of Europe, Chili, and even Australia, and unforeseen differences from the fauna of the Atlantic States. To such facts, bearing upon the geographical distribution of insects, I pay especial attention in the introductory paragraphs to each family; and, at the end, I give a general survey of the results obtained.

For the fauna of Colorado, I availed myself of very valuable materials kindly contributed by Mr. P. R. Uhler, Dr. A. S. Packard, and

Lieut. W. L. Carpenter. Here and there I have introduced descriptions of some remarkable species from the Atlantic States.

In treating of the Californian fauna (or flora) it must be borne in mind that what is called Sierra Nevada is not only a mountain range, but a whole country—a high plateau from 6,000 to 8,000 feet above sea-level, forming a long and comparatively broad belt of land, with its lakes, rivers, forests, and plains—an upper story of California, partaking of some of its products, but on the whole entirely different. For the better undestanding of the facts bearing on the geographical distribution of insects, I will state here, once for all, that my collections about Summit, Sierra Nevada, and Webber Lake were formed at an altitude from 7,000 to 8,000 feet above sea-level; that the altitude of Lake Tahoe is 6,200 to 6,300 feet, and that of Yosemite Valley about 4,000 feet. My collecting grounds in Southern California, as well as in Marin and Sonoma Counties, were all at comparatively low levels, except the Geysers, Sonoma County, which are about 3,000 feet above sea level.

It is not my intention to describe *all* the western *Diptera* which I possess or can get hold of. Always keeping the higher aims of science in view, my effort will be to contribute toward those aims. The detailed description of special entomological faunas must of necessity be left to local students. An outsider, a transient collector and describer, has to keep their interest in view, and to try to pave the way for them rather than to block up their progress by an indiscriminate and aimless publication of new species.

In prefixing diagnoses to some of my descriptions, my aim was to enable the reader at a single glance to get hold of the principal features of the described species, and thus to save his time in the work of identification. Such a diagnosis, in order to be useful, must be short, even at the risk of being applicable to more than one species. Wherever the species in a genus are more numerous, I prefer to give an analytical table. The attempt of some authors to draw diagnoses which are tantamount to definitions of the species is very difficult to carry out, especially in the larger genera; such diagnoses finally become as long as the descriptions themselves, and therefore practically useless.

In quoting species described in North American publications or in Dr. Loew's "Centuries", I will simply refer to them without repeating the descriptions, as it is to be expected that a dipterologist is in possession of the works thus quoted. In some cases I will reproduce or translate descriptions which are less easily accessible.

All the type specimens of these my papers I intend to deposit, for future reference, in the Museum of Comparative Zoölogy in Cambridge, Mass., where my former dipterological collections are also to be found ; the few exceptions will be mentioned in their place.

I owe a special tribute of gratitude to Mr. Henry Edwards, of San Francisco, for his manifold assistance, as well as for the contribution of valuable specimens.

Families CULICIDÆ, CHIRONOMIDÆ, PSYCHODIDÆ.

Half a dozen species of *Culex*, two *Anopheles*, and two *Chironomus* are among my collections from California. They all exhibit the characters and coloring peculiar to the species of these genera in other countries. A *Culex* from Southern California is distinguished by very sparsely bearded antennæ of the male and a peculiar structure of the palpi.

PSYCHODA sp.—A single specimen; San Rafael, Cal.

In the absence of any remarkable western forms, I describe two new species from the Atlantic States. The first belongs to the little known genus *Aëdes* (*Culicidæ*), of which only one species was known to occur in the United States. The other is a second species of the new genus *Chasmatonotus* (*Chironomidæ*) established by Dr. Loew for a species which I discovered in the White Mountains.

AËDES FUSCUS n. sp., ♂ ♀.—Brown; thorax clothed with a short, appressed, brownish-golden tomentum; abdomen with whitish-yellow narrow bands at the base of the segments; venter whitish-yellow. Antennæ black; proboscis and legs brownish, with a metallic reflection; femora paler on the under side; pleuræ under the root of the wings with a spot clothed with whitish scales. Long. corp. 3–4^{mm}.

Hab.—Cambridge, Mass., in May.

Obs.—I bred this species from larvæ which I found in a pool together with those of several species of *Culex*. The larvæ and pupæ behaved exactly like those of *Culex*, and only attracted my attention by their smaller size. If I could have known beforehand that they belonged to *Aëdes*, I would have compared them more closely with the larvæ of *Culex*. The metamorphosis of *Aëdes* has never been observed before.

CHASMATONOTUS BIMACULATUS n. sp., ♂.—Black; wings of the same color and with two large white spots. Length about 1.5^{mm}.

Black; thorax shining; base of the abdomen laterally pale greenish-yellow. Feet black; front coxæ and base of all the femora yellowish; the first tarsal joints are of the same pale yellowish color, except the tip, which is black. Knob of halteres greenish. Wings black; the first white spot is in the shape of a cross-band between the second vein and the anal angle; the second spot is square, and situated on the hind margin, within the fork of the fifth vein.

Hab.—Catskill Mountain House, in July, 1874; numerous male specimens; Quebec (Mr. Bélanger).

The first posterior cell and the cell within the fork of the fifth vein are much longer here than in *C. unimaculatus* Lw., and the latter cell is larger and broader. Hence it happens that although in both species the cross-band-like spot is placed immediately inside of the proximal end of the fork, it occupies the middle of the wing in *C. unimaculatus*, and is much nearer the base in *C. bimaculatus*. The abdomen of the male ends in a comparatively large and conspicuous forceps (the "*hypopygium maris globosum*" in Mr. Loew's description of *C. unimacu-*

latus seems to indicate a different structure ?). I found both species in the same situation, walking in numbers on the leaves of low shrubs.

Family CECIDOMYIDÆ.

Of the numerous galls of *Cecidomyiæ* observed by me in California, I will mention only a few, of which I have kept a written record.

On *Juniperus californicus*, fleshy, subglobular galls on the axis of the small twigs; when full grown, about two-fifths of an inch in diameter, with a round opening at the top, the edge of which is from three- to five-lobed, the gall when ripe thus resembling the fruit of the Medlar (*Mespilus*) in shape; but, before being full grown and open, it is more like a diminutive melon or tomato, being furrowed longitudinally, like these fruits. The furrows are usually six, probably representing six leaves round the axis of the plant. At the base of the gall, round its attachment, there are three sepal-like, small, fleshy, bilobed leaflets. The reddish larva in the cavity of the gall is smooth, and shows no vestige of a breast-bone; in more mature galls, the pupa, glued to the bottom of the cavity, could be distinctly seen through the opening at the top. Very common in March, 1876, about Crafton's Retreat, twelve miles from San Bernardino, Cal.

On *Lupinus albifrons;* folded leaves, forming a pod-shaped swelling; each contained several larvæ, inclosed in a delicate cocoon. Very common about Lone Mountain, San Francisco, in April.

On *Audibertia* sp. (*Compositæ*); swelling on leaves and leaf-stalks, with a neck-shaped prologation, open at the top, the whole having the shape of a round-bellied bottle; sometimes two or three such bottles, alongside of each other, coalescent; inside a longitudinal canal, at the bottom of which I found in several instances a pupa of *Cecidomyia;* wings and thorax blackish; abdomen red; no horny projections anteriorly. A small Hymenopterous parasite often infests this gall. Santa Barbara, end of January, and later in other localities; not rare.

On *Garrya fremonti*, succulent, green swellings on male flowers, contain larvæ and pupæ apparently of a species of *Asphondylia*. On the heights about Yosemite Valley, at an altitude of 7,000 to 8,000 feet, in June.

On *Artemisia californica* (?), accumulation of leaves, produced by the arrested growth of lateral shoots. About Los Angeles, Cal. Inside I found pupæ of *Cecidomyia*, nearly ripe, on the 3d of March.

On *Baccharis pilularis* (syn. *sanguinea*), rounded accumulation of deformed and swollen leaves at the end of twigs; contains larvæ of *Cecidomyia*, from which I bred the fly.

Family MYCETOPHILIDÆ.

Seems abundantly represented in California, although I did not collect very diligently in it. Among my few specimens, I find the following genera:—

PLATYURA sp.—San Rafael, April 12; venation like tab. xix, f. 7a, of the Monograph of the European *Mycetophilidæ* by Winnertz.

PLATYURA sp.—Fossville, Napa County, Cal., May 7. Large red species, with the apex of the wing and a central cloud brown; the anterior branch of the second vein connects it, in the shape of a cross-vein, with the latter part of the first vein.

BOLETINA sp.—Yosemite Valley.

SCIOPHILA, 2 species.

DOCOSIA sp.—Yosemite Valley, June 8; venation exactly like Winnertz's tab. xx, f. 23a.

MYCETOPHILA sp.—San Rafael, Cal., April. Of the group of the European *M. lunata*, and very like it.

EXECHIA sp.—Yosemite Valley.

GNORISTE MEGARRHINA n. sp.—Proboscis nearly as long as the body, filiform. Length of the body, 7^{mm}; of the proboscis 5.5^{mm}; face deep velvet-black, opaque; antennæ brown, second joint somewhat reddish; proboscis brown; vertex black, with a slight gray pollen; thorax brownish-yellow, with three black stripes on the dorsum, the intermediate geminate; halteres pale yellow; legs yellow; tarsi infuscated; wings with a slight yellowish tinge; a light gray shadow along the hind margin, beginning at the apex.

Hab.—Yosemite Valley, June 10.—One specimen.

Although the proboscis of this species is much longer than that of the European *G. apicalis*, they agree in all essential characters, and there is no necessity for establishing a new genus. *G. megarrhina* has the venation of *G. apicalis* (Winnertz, l. c., tab. xx, f. 16); only the proximal end of the fork of the fifth vein is a little nearer to the root of the wing, and the costa is prolonged a little beyond the tip of the second vein.

Family BLEPHAROCERIDÆ.

The new species which I describe is the tenth now known species of this remarkable family,—remarkable for its exceptional characters; for the paucity of the species, scattered through the most distant parts of the world; and for the variety of generic modifications which these species show in preserving at the same time with wonderful uniformity the very striking family characters, some of which are unique in the whole order of *Diptera*. Among those ten species, three belong to the United States; one I found abundantly in a locality near Washington, D. C.; the second was discovered by Lieut. W. L. Carpenter in the Rocky Mountains; the third, described below, comes from Yosemite Valley. A list of the known species of the family, in chronological order of publication, with the locality of each, may find its place here:—

Blepharocera fasciata (Westw.), in Guérin-Méneville, Magaz. de Zool., 1842.—Albania, in Europe.

Liponeura cinerascens Loew, Stett. Entom. Zeit., 1844.—Europe.

Apistomyia elegans Bigot, Ann. Soc. Entom. de France, 1862.—Corsica

Blepharocera capitata Loew, Centur., iv, 1863.—District of Columbia.
Paltostoma superbiens Schiner, Verh. zool.-bot. Ges., 1866.—Colombia, South America.
Liponeura bilobata Loew, Bullet. Soc. Entom. Ital., 1869.—Southern Italy and islands of Greece.
Hammatorhina bella Loew, Bull. Soc. Entom. Ital., 1869.—Ceylon.
Bibiocephala grandis Osten Sacken, in Dr. Hayden's Geol. Rept. for 1873.—Rocky Mountains.
Hapalothrix lugubris Loew, Deutsche Ent. Mon., Berl., 1876, p. 213.—Monte Rosa (Italian side).

Blepharocera yosemite n. sp.

Blepharocera yosemite is closely allied to the known species, both of the genera *Blepharocera* and *Liponeura*. The differences it shows, although important, do not necessitate the immediate formation of a new genus for it, the more so as sooner or later new additions to the family *Blepharoceridæ* will probably require a remodeling of the now adopted genera.

The structural characters of the species are as follows :—

Eyes pubescent, separated by a moderately broad front; upper smaller portion of the eye with large, lower larger portion with small, facets. *Antennæ* 14 jointed, about twice as long as the head, and of equal breadth, that is, not tapering toward the end ; first joint very short and small, the second a little larger, the third long, cylindrical, equal to the two following taken together, the fourth and following joints subcylindrical, attenuated at the base. *Legs* long and comparatively strong ; a large and stout spur at the end of the hind tibiæ; a much smaller spur alongside of it; ungues with a tooth-like incrassation at the base. *Wings* comparatively larger and broader than in *Blepharocera ;* anal lobe very large, projecting. *Venation :* second submarginal cell short and petiolate, the petiole being about equal in length to the interrupted vein between the incomplete second and third posterior cells (in other words, the third vein does not issue near the small cross-vein, but from the second vein, at a distance from the small cross-vein, about equal to the abbreviated vein). Between the base of the fourth posterior cell and the preceding (fourth) longitudinal vein, a cross-vein exists (as it does in *Liponeura bilobata* and in *Bibiocephala*). In other respects, the venation resembles that of *Blepharocera* and *Liponeura*. *Forceps* of the male large, its lobes flattened, as if coriaceous (even in the living insect).

It follows from this enumeration that in the structure of the front the present species is nearer to *Liponeura*, the eyes of *Blepharocera* being subcontiguous ; in the structure of the facets of the eyes, it is like *Blepharocera* and unlike *Liponeura*, where the facets are said to be of equal size on both halves of the eye. From both genera it differs in the shortness of the second submarginal cell. It resembles *Liponeura bilobata* in the presence of a cross-vein between the fourth vein and the fork behind it, a cross-vein which is wanting in *L. cinerea* and in *Blepharocera*.

The antennæ have one joint less than those of *Blepharocera* (I counted them on the living specimen), and although proportionally of the same length, they are not subsetaceous, as in the latter genus, and have much more distinctly marked joints.

BLEPHAROCERA YOSEMITE n. sp., ♂.—Body brownish-gray; wings tinged with brown, their distal third hyaline. Length 6–7^{mm}; wing 9^{mm}.

Body brownish; thorax above with a grayish pollen, abdominal incisures slightly whitish, more distinctly so on the sides of the venter; genitals reddish; antennæ brownish, paler at base; legs yellowish-brown; the tips of the femora infuscated; wings tinged with brown, this brown with a distinct bluish opalescence; distal third of the wings hyaline.

Three male specimens caught by me on the wing, on the bridle-path to the foot of the Upper Yosemite Fall, June 6, 1876, about 3 p. m.

Family TIPULIDÆ.

The enumeration which I give contains some thirty-five species from California, belonging to the first six sections of the *Tipulidæ*, commonly united under the name of *Tip. brevipalpi*,—a comparatively small number, considering that, owing to my early studies in this family, I paid more attention to it perhaps than to any other. The paucity of *Eriopterina* was especially striking. *Trichocera*, which one would naturally expect during the warm winter days of that climate, did not appear at all; I found a single specimen of a rather peculiar species later in the spring.

Among these thirty-five species, seventeen are identical with species from the Atlantic States, or at least so closely resembling them as to be provisionally classed among the species of doubtful identity. Two of that class of species are at the same time European,—*Symplecta punctipennis* and *Trimicra pilipes*. The very common occurrence of the latter all over California during winter and spring is worthy of notice.

Most of the species peculiar to California belong to genera represented in other parts of the world:—*Dicranomyia* (2 sp., one of which undescribed); *Limnobia* (2 sp.); *Erioptera* (2 sp.); *Elliptera* (1 sp.); *Goniomyia* (1 sp., undescribed); *Limnophila* (4 sp., only one described); *Trichocera* (1 sp.); *Amalopis* (1 sp.); *Pedicia* (1 sp.); *Eriocera* (1 sp.). Among these, the following deserve to be noticed:—

Elliptera, a genus belonging to the remarkable and intermediate group *Limnobina anomala*, was among the few European genera which have not hitherto been discovered in North America. I found a number of specimens in the Yosemite Valley, which reproduce exactly the generic characters of *Elliptera*, although they belong to a species different from the only European species hitherto described.

Eriocera californica belongs to the *Erioceræ* with very long antennæ in the male, of which three species occur in the Atlantic States, one in

Chili, and two fossil species have been found in the Prussian amber. I am not aware of such species having been found in other countries, although *Eriocerœ* with short antennæ in both sexes are everywhere abundant in the tropics.

Pedicia is represented by a single species, analogous to the Eastern American and the European species, but different from both.

The new genus *Phyllolabis*, with two species, is peculiar to California, and remarkable for the large development of the forceps of the male.

Of the two sections intermediate between the *Tipulidæ brevipalpi* and *longipalpi*, no *Cylindrotomina* have as yet been discovered in the western region. The *Ptychopterina* are represented by two species:—

Ptychoptera lenis n. sp., which belongs to the whole western region from California to Colorado.

Protoplasta vipio n. sp., perhaps the most interesting of all the Californian *Tipulidæ*, closely allied to the Chilian *Tanyderus*, the fossil amber-genus *Macrochile*, and the North American *Protoplasta fitchi*.

Bittacomorpha has not as yet been found in California, but *B. clavipes* occurs in Oregon.

The *Tipulidæ longipalpi*, in contrast to the *brevipalpi*, are very abundantly represented in California, both in the number of species and of specimens. The larvæ probably live on the roots of the rich and abundant Californian grasses. I have abstained from working up this part of my collection, owing to the large number of closely allied species and my insufficient knowledge of the *Tipulidæ* of the Atlantic States.

The gigantic *Holorusia rubiginosa* is a peculiar Californian form. However, Dr. Loew, in establishing the genus, mentions *Holorusiæ* from Java (Centur., iv, 1); elsewhere he describes one from the island Bourbon.

Pachyrrhinœ are much rarer in California than in the Atlantic States.

A species of *Dixa* occurs in California; but I have only a single imperfect specimen (San Geronimo, Marin County, April 19).

Section I.—*Limnobina.*

GERANOMYIA CANADENSIS (Westwood), Osten Sacken, Monogr., iv, p. 80.—Male and female from Los Angeles, February. A common species in the Atlantic States.

DICRANOMYIA BADIA (Walker), Osten Sacken, Monogr., iv, p. 72.— Common in the Atlantic States. San Rafael, Cal., March 31, April 13.

DICRANOMYIA DEFUNCTA Osten Sacken, Monogr., iv, p. 76.—Common in the Atlantic States near springs or water running over dams. Santa Cruz, Cal., May 21, three males in the same situation. I observed the structure of the forceps, peculiar to this species, on the specimens when they were still alive. A single specimen from Webber Lake, July 24, has the wings much less densely spotted, and with a cross-vein in the submarginal cell. The cross-vein, however, may be merely adventitious.

DICRANOMYIA MARMORATA Osten Sacken, Monogr., iv, p. 77.—A Californian species. I found a male and a female near Saucelito, Cal., April 2. In the live insect, I noticed a peculiarity, which I had over-looked in the dry ones, from which I drew my description. The an-tennæ are distinctly submoniliform, the nearly globular joints being separated by very short pedicels. In my description, the words "re-lated to *humidicola* O. S." better be struck out.

DICRANOMYIA n. sp.—Seems common in Marin County, California, in April. In looking for it on my analytical table (l. c., p. 61,) *D. liberta* and *hæretica* would be reached; it is neither of them, but a new species, which I leave to others to describe, as my specimens are not well pre-served enough for that purpose. The structure of the male forceps will have to be observed in the live specimens.

LIMNOBIA SCIOPHILA n. sp.—Marginal cross-vein some distance back of the tip of the first longitudinal vein; femora with three brown rings; wings with grayish clouds and intervening subhyaline spaces in all the cells; length 10–11^{mm}.

Rostrum, palpi, and antennæ brown, the latter with long verticils; thoracic dorsum with three brown stripes, the intervening spaces, shoulders, middle of the mesonotum, etc., grayish-pruinose; abdo-men brown, incisures paler; genitals yellowish-brown; halteres with brown knobs; femora pale yellow, with three brown rings on the distal half, the last of them very near the tip; tibiæ and tarsi yellow-ish-brown. Wings with a faint yellowish tinge as a ground-color; grayish clouds of irregular shape occupy all the cells, and become almost confluent on the distal half of the wing, leaving only small spaces of the ground-color at both ends of the cells; in four or five places along the first vein, the clouds are darker, so as to have the appearance of brown spots; the marginal cross-vein is in the middle of the stigma, and some distance back of the tip of the first vein.

Hab.—Marin and Sonoma Counties, California, in the spring; com-mon, especially in dark, deep gulches, with running water at the bottom (Menlo Park, March 25; San Rafael, April, May; Geysers, Sonoma County, May). Three males and four females.

Very closely allied to the European *L. nubeculosa.*

LIMNOBIA CALIFORNICA Osten Sacken, Monogr., iv, p. 96.—California.

Section II.—*Limnobina anomala.*

DICRANOPTYCHA SOBRINA Osten Sacken, Monogr., iv, p. 118.—A spe-cies very similar to this eastern one, and perhaps identical with it, occurs quite commonly in Marin and Sonoma Counties, California. The two basal joints of the brown antennæ are yellow and the fringe of hairs on the anterior margin of the wings in the male is not very long and conspicuous; in both respects, these specimens are more like the form which I called *D. sororcula* in my former essay, and which later I gave up as a species, perhaps erroneously. The specific characters in this genus require a closer study than I have been able to give them in preparing my Monograph. A male specimen from Lake Tahoe, July 19, is much paler in coloring, and may be a different species.

ELLIPTERA CLAUSA n. sp.—This is an interesting discovery, as the genus *Elliptera* (compare Monogr., iv, p. 122, tab. i, f. 10), represented by a single species in Europe, had not been found in America before. The venation is like that represented on the above quoted figure, only the first longitudinal vein is a little shorter, so that the segment of the margin between its tip and the tip of the second vein is much longer than the segment between the second and third veins (and not shorter, as the figure has it); *the discal cell is closed.* But the characteristic mark of *Elliptera,* the close proximity between the first and second veins, exists also in this new species. *Elliptera* has no empodia and no vestige of a marginal cross-vein. The forceps of the male, which I observed in life, resembles that of an ordinary *Limnophila,* and not at all that of a *Dicranomyia.*

Male and female.—Antennæ and palpi black; front grayish-pruinose; thorax grayish-pruinose; three distinct broad brown stripes on the dorsum; halteres brown, their root yellow; abdomen grayish-brown; legs brown; coxæ and root of the femora, especially of the front pair, yellowish; wings subhyaline, slightly tinged with grayish; stigma oval, brown.

Hab.—Yosemite Valley, Cal. I found umerous specimens on the wet mos, in the spray of the Vernal Fall, June 11. I have now four males and two females before me.

Section III.—*Eriopterina.*

ERIOPTERA DULCIS n. sp.—The præfurca ends in the second submarginal cell; discal cell closed; inner end of third posterior cell much nearer to the root of the wing than the inner end of second posterior; wings pale brownish, with a number of white spots, especially along the margin and on the cross-veins; femora with a dark brown ring before the tip. Length about 3^{mm}.

Thorax yellowish, with brown lines on dorsum and pleuræ; abdomen brownish, halteres with a brown knob; wings pale brown, with numerous white spots, one at the extreme proximal end of the basal cells, with a smaller spot, alongside of it, near the costa; a large square spot between the costa and the fourth vein, covering the origin of the præfurca; a similar spot between the costa and the middle of the præfurca; an oblong spot near the end of the præfurca; another one between the end of the auxiliary vein and the second submarginal cell; rounded spots along the whole margin at the end of all veins except the third; often one or two spots in the middle of the posterior branch of the second vein; the distal end of the four posterior cells likewise spotted. Legs pale yellow, a ring before the end of the femora and the tip of the tibiæ dark brown.

Hab.—Lake Tahoe, Sierra Nevada, California, July 19. Six males.

E. dulcis has a striking resemblance to *E. caloptera* Say of the Atlantic States. An obvious difference between them lies in the structure of the discal cell, which in *E. caloptera* is formed by the forking of the anterior, in *E. dulcis* of the posterior, branch of the fourth vein. The distribution of the white spots on the wings is different in both species, as *E. caloptera* has many spots inside of the cells, while *E. dulcis* has none· Besides the brown ring at the tip of the femora, *E. caloptera* has a second one about the middle. The discal cell here is generally closed, while in *E. caloptera* it is more often open. Nevertheless, the homologies between the present species and the group *Mesocyphona*, to which *E. caloptera* belongs, are very striking. They consist in the position of the brown thoracic stripes; in the presence of a brown ring on the femora; as far as I can see, in the structure of the male forceps, which resembles the figure I gave of the forceps of *E. caloptera* (l. c., tab. iv, f. 15); and the course of the last longitudinal vein, which is undulating, and at the same time diverging from the preceding vein, thus holding the middle between the converging arcuated seventh vein of the subgenus *Erioptera* and the straight and diverging one of the subgenus *Acyphona*.

Thus, *E. dulcis* would be well placed in the same subdivision with *E. caloptera*, the subgenus *Mesocyphona;* only the definition of the subgenus should be modified, and less stress laid on the forking of the anterior or posterior branches of the fourth vein. The subdivisions I proposed for the genus *Erioptera* (Monogr., iv, p. 151) were based mainly on the sixteen North American species which I knew at that time. I believe that in the main they will hold good in a more general application, only their definitions will have to be modified in some points, and based upon a closer study of the male forceps.

ERIOPTERA BIPARTITA n. sp.—The præfurca ends in the second submarginal cell; the anterior branch of the fourth vein is forked, and by means of two cross-veins forms a double discal cell; wings spotted with brown along the margin and on the cross-veins and forks. Length 3.5–5^{mm}.

Male and female.—The wings of this species are exactly like those of my *E. graphica* of the Atlantic States (Monogr., iv, Tab. i, f. 18); only the stump of a vein, which in that species, as the figure shows, protrudes inside of the discal cell, is prolonged here, so as to reach the anterior branch of the fourth vein, and to form a fork with it. The two cross-veins in the second and third posterior cells thus inclose two disca cells. The distribution of the brown spots is the same as on the above-quoted figure, with some slight differences: the spot at the root of the præfurca crosses the subcostal cell and reaches the costa; that at the end of the first vein is smaller, and stops short before crossing the first submarginal cell; the seventh vein about its middle has a stump of a vein, projecting into the anal angle, and that is also marked with a brown spot; the fifth vein is checkered with brown spots; thorax yellowish-gray, with an indistinct double brown stripe in the middle; halteres yellowish; abdomen brown; legs yellow, tips of the femora and of the tibiæ slightly infuscated; male forceps with strong horny black appendages.

Hab.—San Rafael, Cal., April, May; San Francisco, June. Two males and a female. In one of the males, the cross-veins in the second and third posterior cells are wanting, so that both discal cells are open.

ERIOPTERA FORCIPULA Osten Sacken, Monogr., iv, p. 163.—I have a male and two females from San Rafael, Cal., March 31, and Lagunitas Creek, Marin County, April 15, which agree very well with the specimens from the Atlantic States. The male has an uncommonly large forceps, of complicated structure; on the comparison of the detail of that structure, impossible in dried specimens, would rest the final identification of the species.

E. forcipula belongs to the subgenus *Molophilus.* I have two other Californian species of the same subgenus, but only a single specimen of each (San Rafael, March 31; Summit Station, Sierra Nevada, July 17).

(?) ERIOPTERA URSINA Osten Sacken, Monogr., iv, p. 164, of the Atlantic States is perhaps identical with the European *E. murina.* These singular, small, black flies form swarms above running waters in shady places. I have observed the same phenomenon on Lagunitas Creek, Marin County, April 14, but have kept only a single, badly preserved specimen.

TRIMICRA PILIPES (Fabricius), a European species, the description of which may be found in Schiner, Fauna Austr., Diptera, ii, p. 536; a more detailed one in Schummel, Beitr. z. Ent., p. 152 (*Limnobia fimbriata*). About the genus *Trimicra*, established by me, see the Monographs, iv, p. 165, tab. ii, f. 1, wing.—This very common Californian species, occurring everywhere through the winter and spring, I hold to be identical with the European species (I have specimens from Angel Island, January 11; Santa Barbara and Los Angeles in February; San Rafael in April; Santa Cruz in May). The specimens differ very remarkably in size, the largest measuring up to 8^{mm}; the smaller specimens are usually females. In identifying these specimens with the European species, I rely upon the descriptions of the latter and my recollection of them; I have no specimens for comparison. *Trimicra anomala* O. S. of the Atlantic States is much more rare, and I have never found it as large as the other. Nevertheless, I think now that this also is the same species. The specimens which I have seen from Mexico and South America also resemble *Trimicra pilipes* very much (compare Monogr., iv, p. 167), and it is not at all improbable that this species, like the following, its close relative, has a very wide distribution.

SYMPLECTA PUNCTIPENNIS, common in Europe and in the Atlantic States, occurs in Colorado and is common in California. I have specimens from Los Angeles and Santa Monica, taken in February; from San Rafael, March 31; Lake Tahoe, July 18. I have observed before (Monogr., iv, p. 171) that *Idioneura macroptera* Philippi from Chili is

probably *S. punctipennis*. On the figure of the wing which I gave (l. c.,
tab. i, f. 20), the brown spot at the base of the præfurca is nearly invisible; this was an individual peculiarity of the figured specimen;
usually it is much larger.

GONIOMYIA sp.—A male specime from Lake Tahoe, July 19, is very
like *G. subcinerea*, especially in the venation (Monogr., iv, tab. ii, f. 4),
only the legs are darker, the halteres decidedly brown, the pleuræ with
a distinct brown stripe, which is wanting in *G. subcinerea*. A female
from Saucelito, Marin County, Cal., April 2, is comparatively larger,
and has the wings slightly tinged with brownish.

Section IV.—*Limnophilina.*

LIMNOPHILA TENUIPES Say, Osten Sacken, Monogr., iv, p. 210.—I
have two females from Lake Tahoe, July 19, which I cannot distinguish
from this species.

LIMNOPHILA LUTEIPENNIS Osten Sacken, Monogr., iv, p. 217.—Found
abundantly near San Bernardino, Cal., in March. I cannot find any
difference between these specimens and eastern ones.

LIMNOPHILA APRILINA Osten Sacken, Monogr., iv, p. 223.—A male
from Summit Station, Sierra Nevada, July 17, does not show any perceptible difference from eastern specimens. A male and a female from
San Rafael, Marin County, Cal., April 13, have very dark brown and
well marked thoracic stripes; the coloration of the wings is much
darker, the spots along the anterior margin are larger; that at the end
of the first longitudinal vein, for instance, almost coalesces with the
brown cloud at the base of the first posterior cell. These specimens
may perhaps be considered a different species.

LIMNOPHILA MONTANA Osten Sacken, Monogr., iv, p. 227.—Two males
and a female from the Geysers, Sonoma County, California, May 5-7.
The specimens have become somewhat greasy, so that the coloring cannot be compared with that of the eastern specimens; the wings are exactly the same. The male forceps shows the peculiar structure of the
subgenus *Dactylolabis*, to which the species belongs.

LIMNOPHILA MUNDA Osten Sacken, Monogr., iv, p. 226.—Specimens
from San Rafael, Cal., May 14, and Sonoma County, May 5-7, resemble
this eastern species very much, but require a closer comparison.

LIMNOPHILA ADUSTA Osten Sacken, Monogr., iv, p 215.—Two specimens from Webber Lake, Sierra County, seem to belong to this species,
or rather group of species.

LIMNOPHILA DAMULA n. sp.—Gray; antennæ black; wings spotted
with brown. Length 6-7^{mm}.

Rostrum, antennæ, and palpi black; thorax gray, with faint brown
stripes; halteres whitish; abdomen brownish-yellow in the male, nearly
brown in the female; ovipositor yellowish; legs yellowish-brown. Venation like that of a *Dactylolabis* (Monogr., iv, tab. ii, f. 7); that is,
the first submarginal cell long, rather angular at the proximal end, its

petiole short; the marginal cross-vein is inserted at the end of the first vein a little before the middle of the first submarginal cell; second submarginal cell but a trifle longer than the first posterior cell; five posterior cells. Coloring of the wing whitish-hyaline; a brown spot fills out the proximal end of the first basal cell; another one, inverted T-shaped, at the root of the præfurca; brown clouds at the proximal end of the first submarginal cell, on the central cross-vein and on the great cross-vein; smaller clouds at the base of the second and third posterior cells; stigma pale brown, with a brown cloud on the marginal cross-vein; a pale cloud at the end of the seventh vein; in most specimens, a few brownish dots are scattered in the areas of the cells irregularly here and there. The size and intensity of the spots on the wings vary in different specimens.

Hab.—Crafton, near San Bernardino, Cal., in March, not rare. Two males and two females.

The forceps of the male shows the digitiform appendages peculiar to the subgenus *Dactylolabis* (Monogr., iv, tab. iv, 26); a second forceps-like organ, slender, horny, is visible below them. The ovipositor of the female differs from that of any Tipulid I know of; the horny plate, usually existing at the base of the ovipositor, is so prolonged here as to cover and conceal this organ. The end of this long plate, with parallel sides, is split in the middle, and produced on each side of the cleft, in a small, curved point, diverging from the corresponding point on the other side. This end of the horny plate is yellowish, the basal portion being dark brown. On the under side, the plate is hollow, canaliculate, and contains, some distance before its end, the small ovipositor, which is thus entirely invisible from above.

I have three more species of *Limnophila*, which seem peculiar to California, but only in single specimens. For this reason, I abstain from describing them.

PHYLLOLABIS nov. gen.

Two submarginal cells; four posterior cells; discal cell closed; subcostal cross-vein a very short distance before the tip of the auxiliary vein, which is immediately before the stigma; no marginal cross-vein; first submarginal cell about half as long as the second, its slightly arcuated petiole occupying the length of the other half; the second vein and both of its branches are pubescent; the other veins are somewhat pubescent before their ends; eyes glabrous; antennæ 16-jointed; tibiæ with exceedingly small but distinct spurs at the tip; empodia small but distinct; ungues smooth. The abdominal segment bearing the genitals is unusually swollen in both sexes, bearing in the male a large forceps with horny appendages on the inner, and a long foliaceous whitish appendage on the under side. Belongs to the *Limnophilina* with four posterior cells, although, owing to the unusual structure of the male genitals and the total absence of a marginal cross-vein, its immediate relationship to the other genera of the group is not apparent.

The stature and general appearance are those of a *Limnophila*.

Antennæ, when bent backward, nearly reach the root of the wings; the joints of the scapus have the usual structure; the flagellum is not perceptibly stouter at the base than at the end; the joints have that subcylindrical shape, attenuated at the point of insertion and verticil-bearing about the middle, which is quite common among the *Limnophilina*; they gradually decrease in length toward the end and become more oval; verticils short; front moderately broad. *Vertex* but moderately convex.

The *legs* are long and slender, with an almost imperceptible pubescence; the spurs at the end of the tibiæ are very distinctly seen under a magnifying power of 100 to 150. The ungues and empodia are exceedingly small.

The *wings* are of a moderate length and breadth; the venation has been partly described above; the stigma is well defined, oval, placed at the end of the first vein. The præfurca has very little curvature at the base, and is not much longer than the petiole of the fork of the second vein; the second submarginal and first posterior cells are of equal length, their bases being nearly on the same line; the sides of the first posterior are almost parallel; the structure of the discal cell shows that it is formed by the forking of the posterior branch of the fourth vein, connected by a cross-vein with the anterior branch; the great cross-vein is at the bifurcation of that posterior branch, and thus a little beyond the middle of the discal cell.

The genitals of the male are very large and club-shaped, resembling those of a *Tipula* more than those of a *Limnophila*. The following is the description of the forceps of *P. pallida*, as I do not possess the male of the other species :—

The last upper abdominal half-segment is uncommonly large and convex; two large basal pieces of the usual shape, bearing a small, curved, pointed rostriform appendage at the end, and some branched and hairy inner appendages; on the under side of the forceps, and entirely detached from it, is a very characteristic yellowish-white elongated foliaceous appendage, folded lengthwise and bifid at the tip.

The female has the last abdominal segment likewise distinctly swollen, especially when seen from the side.

Stigma pale brownish; antennæ uniformly black .... *claviger*.
Stigma dark brown; two basal joints of the antennæ pale
   brownish; the rest black ....*encausta*.

PHYLLOLABIS CLAVIGER n. sp.—Head gray, with a shade of brownish on the front; antennæ and palpi black. Thorax gray, with three ill-defined brownish stripes, the intermediate subgeminate; halteres yellowish-white; knob sometimes faintly infuscated. Abdomen grayish-brown; male forceps brown; foliaceous appendages yellowish-white. Legs brown; coxæ and base of femora brownish-yellow. Wings gray-

2 H B

ish-hyaline; stigma oblong, pale brownish-gray; subcostal cross-vein a little distance back of the tip of the auxiliary vein.

Hab.—Crafton's Retreat, near San Bernardino, Cal., in March. Three males and six females.

The ovipositor of the female has nearly straight, ferruginous valves, smooth on the under side. Length 6–7ᵐᵐ.

PHYLLOLABIS ENCAUSTA n. sp.—Head gray; antennæ brown, two basal joints yellowish; palpi brown. Thorax gray, the dorsum with a brownish-yellow tinge; knob of halteres infuscated. Abdomen brownish-black; in the female, the two last segments are remarkably turgid above and below; upper valves of the ovipositor short, curved, distinctly serrate on the under side. Legs brownish-yellow, tarsi darker toward the end. Wings subhyaline; veins brown, except at base and near the costa, where they are pale yellow; subcostal cross-vein close by the tip of the auxiliary vein. Length 6–7ᵐᵐ.

Hab.—Lagunitas Creek, Marin County, California, April 15–20; San Mateo, Cal., April 9. Four females.

TRICHOCERA TRICHOPTERA n. sp.—Antennæ brown, second joint slightly paler; thorax dull grayish-pollinose, without any distinct stripes; halteres with a brown knob; abdomen brown above, yellowish below; wings subhyaline, immaculate; veins very distinctly pubescent; venation in the main like Monogr., iv, tab. ii, f. 13, but the discal cell smaller; posterior cells 2–4 longer; the cross-vein at the proximal end of the fourth posterior cell is placed obliquely, and thus does not correspond with its posterior end with the great cross-vein (this may, however, be merely adventitious). Feet pale yellowish. Length 3–4ᵐᵐ.

Hab.—Lagunitas Creek, Marin County, California, April 15. One female.

This is the only specimen of the genus *Trichocera* which I came across in California. The species will be easily distinguished by its distinctly pubescent wing-veins.

Section V.—*Anisomerina.*

ERIOCERA CALIFORNICA n. sp.—*Male.*—Antennæ more than twice as long as the body; basal joints reddish; flagellum reddish-brown, beset on its under side with two rows of small spine-like bristles; head reddish above; palpi brown; thorax brownish, with a grayish pollen, especially on the pleuræ; three brown stripes on the dorsum, the intermediate one geminate; a faint brownish stripe on the pleuræ; halteres with a brownish knob; abdomen brown, lateral edges yellowish; genitals reddish; legs reddish; tips of femora and of tarsi brown; wings with a strong brownish-yellow tinge, more saturate in the costal cell; five posterior cells. Length about 15ᵐᵐ.

Hab.—Marin County, California (H. Edwards). Two males.

Belongs to my subgenus *Arrhenica* (Monogr., iv, p. 252), and is closely allied to my *E. spinosa* from New England.

In my Monograph, I omitted to mention that *Megistocera chilensis* Philippi from Chili is very probably not a *Megistocera* at all, but an *Eriocera* with long antennæ. This is the only South American species with long antennæ which I know of as yet, and the circumstance that it belongs to Chili is in keeping with certain other analogies already noticed between the Chilian and the Californian fauna.

ERIOCERA BRACHYCERA n. sp., ♂ ♀.—Antennæ short in both sexes; five posterior cells; thorax brownish-yellow, with four brown stripes; abdomen brown. Length, male, 14^{mm}; female, with ovipositor, 22^{mm}. Antennæ of the male about as long as head and thorax together; basal joints reddish, the remainder brown; the first joint of the flagellum is the longest; the three following but little shorter; the end of the last has the appearance of bearing a seventh minute joint; antennæ of the female shorter than those of the male; the first joint of the flagellum is the longest; the following joints gradually decrease in length and become indistinct; frontal tubercle brownish above, yellowish in front; it bears a tuft of blackish hair. Thorax brownish-yellow, sometimes more grayish above, with four brown stripes; pleuræ brown, with a hoary bloom, which extends over the coxæ. Abdomen brown, but little hairy; valves of the ovipositor but very little curved, long and narrow, ending in a blunt point; halteres reddish, with a brown knob; legs brown; base of femora reddish. Wings strongly tinged with brown; stigma brown, oval; five posterior cells, the second on a long petiole.

Hab.—White Mountains (H. K. Morrison). Two males and a female.

I find now that what I described in the Monogr., iv, p. 253, as the female of *Eriocera spinosa* is the female of the present species.

Section VI.—*Amalopina.*

AMALOPIS CALCAR Osten Sacken, Monogr., iv, p. 268.—A single male, from Webber Lake, Sierra County, July 25, does not show any perceptible difference from the eastern specimens, except that it is somewhat paler in coloring; the male forceps is yellowish and not brownish; the venation is exactly like tab. ii, f. 14, except that both the second submarginal and second posterior cells are petiolate and not subsessile (the same is the case with most specimens of *A. calcar;* the one whose wing is figured happened to be somewhat abnormal).

AMALOPIS n. sp.—One female specimen from Crafton, near San Bernardino, Cal., in March. Body brownish; wing immaculate; venation like l. c., tab. ii, f. 14, except that the præfurca is a little shorter, the second posterior cell petiolate. I do not name it, as I have only a single damaged specimen.

PEDICIA OBTUSA n. sp.—I have seen a single specimen of this species in Mr. Henry Edwards's collection in San Francisco. It was taken near Saucelito, Marin County, Cal., in the spring. Not having the specimen

before me, I cannot give a detailed description; but the following state
ment, prepared from some data kindly furnished by Mr. Edwards in a
letter, will render the species recognizable.

It is very much like both *P. rivosa* and *albivitta*, but differs in the
abdomen being unicolorous, brown above, without the brown dorsal
stripe (consisting of a series of triangles in *P. albivitta*, more continuous
in *P. rivosa*) and the whitish lateral borders. The brown design of the
wings is nearly the same, but the brown is *not* continued toward the
margin, along the last section of the fifth vein; it thus forms an angular stripe along the central cross-veins and the anterior section of the fifth
vein; the hyaline space inclosed between this stripe and the brown an
terior border is smaller and more curved than in *P. albivitta.*

The interruption of the brown stripe along the fifth vein, before reach
ing the margin, occasionally takes place in both *P. rivosa* and *albivitta* ;
at least, I consider *P. contermina* Walker, which shows this peculiarity
as a mere variety of *P. albivitta.*

RHAPHIDOLABIS spec.—A single specimen from Crafton, San Bernar
dino, Cal., has the wings exactly like Monogr., iv, tab. ii, f. 17. The
thorax having become greasy, I am not able to ascertain whether it is
R. tenuipes or not.

List of TIPULIDÆ BREVIPALPI *from Colorado.*

Dicranomyia longipennis.—Europe; Atlantic States; Denver, Colo
(Uhler).

Limnobia indigena.—Atlantic States and Colorado (Kelso's Cabin, foo
of Gray's Peak, at 11,000 to 12,000 feet altitude, July 6, A. S. Packard)

Empeda n. sp.—Georgetown, Colo., July 8 (A. S. Packard).

Erioptera caloptera.—Atlantic States and Colorado.

Symplecta punctipennis.—Europe and North America; also in Chili,
(Boulder City, Colo., June 29, A. S. Packard; Denver, P. R. Uhler.)

Amalopis n. sp.—Idaho, Boulder, Georgetown, Colo., June, July (A
S. Packard). Very like the undescribed species from Southern Cali
fornia, mentioned above; perhaps identical.

Of Section VII, *Cylindrotomina,* no species has been discovered i
California yet, nor in any part of the Western Territories.

Section VIII.—*Ptychopterina.*

PTYCHOPTERA LENIS n. sp.—*Male and female.*—Antennæ black, excep
the first joint, which is red; base of palpi also reddish; hypostom
reddish, in some specimens darker; front black, shining. Thorax black
subopaque; pleuræ silvery-pruinose; scutellum reddish. Abdome
black, shining; male genitals dark brown; ovipositor reddish. Leg
reddish, including coxæ, which are more or less black at the root, an
have a more or less distinct black streak about the middle; hind cox
black, except the tip; femora brown at tip; tibiæ brownish, darker a

tip; tarsi brown, except the base, which is paler. Halteres brownish. Wings with a distinct pale brown tinge, somewhat yellowish along the costa; nearly obsolete clouds, visible with the magnifying-glass only, on the central cross-veins and on the bifurcation of the third vein; venation like *P. rufocincta.* Length, 11–12^{mm}.

Hab.—Yosemite Valley, Cal., June 5; Georgetown, Colo. (8,500 feet altitude), August 12. A male and two females. Seems to resemble the European *P. scutellaris.*

P. metallica Walker from Albany River, Hudson's Bay, may be this species; but Walker states that the venation differs from that of the other *Ptychopteræ,* although it does not appear from his explanation in what the difference consists.

BITTACOMORPHA CLAVIPES (Fabr.).—A specimen from Oregon in Mr. H. Edwards's collection in San Francisco: other specimens were collected in Clear Creek Cañon, Colorado, by Mr. J. D. Putnam, of Davenport, Iowa; near Boulder City, Colorado, by A. S. Packard, June 29.*

Genus PROTOPLASTA.

In my Monograph of the North American Tipulidæ (Monographs of North American Diptera, vol. iv, p. 309), I established two groups of the section *Ptychopterina,* the second of which contains three remarkable and closely allied genera,—*Protoplasa* (better *Protoplasta*), from the Eastern United States; *Tanyderus,* from Chili; and *Macrochile,* a fossil form found in the Prussian amber; all of which with but a single species. These three genera and species differ from the *Ptychopterina* of the first group (*Ptychoptera* and *Bittacomorpha*) in the following characters:—A subcostal cross-vein is present; the second submarginal cell is much longer than the first; the number of posterior cells is raised to six, in consequence of the presence of a supernumerary longitudinal vein in the first posterior cell; collare large.

I have discovered a fourth species in California, which possesses the general characters of the *Ptychopterina,* as well as the particular characters of this second group. The characters enumerated below are merely those in which *P. vipio* differs from the generic characters of *Protoplasta* as given l. c., p. 316. In judging of their importance, it must be borne in mind that the original specimens of *P. fitchi,* at the time when I described them, were old specimens, while I made the description of my new species before the specimen had dried. Some of the minor differences may be due to this circumstance.

Eyes finely pubescent; proboscis together with its large fleshy lips a little longer than the head; antennæ 15-jointed, first joint but little longer than the second; joints of the flagellum elongated, very slightly incrassated on their anterior half, with verticils on the incrassa-

---

* I discover a slip of the pen in my description of this species: Monographs, iv, p. 316, line 2 from top, for *brown,* read *white.*

tion, last joint button-shaped, with a short cylindrical prolongation, which may be taken for a 16th joint. Scutellum projecting. Abdomen of the usual length, ending in the male in a large double-branched forceps. The fourth tarsal joint has a small projection on the under side at the basis (probably a sexual character). Wings narrower than in *P. fitchii*, and anal angle less projecting; venation like l. c., p. 317, f. 7, but the cross-vein in the fifth posterior cell is wanting, and there is no stump of a vein at the origin of the second vein.

P. vipio has the same venation as *Macrochile* Loew (Linn., Entom., v, tab. ii, f. 25), but differs in having 15- and not 19-jointed antennæ. It is also closely allied to *Tanyderus* Philippi (Verh. zool.-bot. Ges., 1865, p. 780). The venation is very like that represented l. c., tab. xxix, f. 57c., only the small cross-vein in the first posterior cell is wanting; the first vein and the branches of the second are more straight, the anal angle more rounded. The neck-like prolongation of the thorax is not quite as long as represented by Philippi. The antennæ of *Tanyderus* are said to be *at least* 25-jointed.

Dr. Philippi's statement that the abdomen of *Tanyderus* ends in two filaments does not warrant his conclusion that the specimen is a male.

Protoplasta vipio has a forceps, each of the two halves of which resembles the thumb and forefinger of a hand when divaricate; that is, each half has two branches, with a deep and broad sinus between them. This forceps, which I observed and sketched from the fresh specimen, has retained its shape after drying. It seems only probable that both *P. fitchi* and *Tanyderus* have a forceps constructed more or less on the same plan, and that the specimens hitherto described were females.

The sexual characters of *P. fitchi* not being as yet known, and *P. vipio* being known in the male sex only, I prefer to leave them provisionally in the same genus, although in the future a generic separation may become necessary.

PROTOPLASTA VIPIO n. sp.—*Male.*—Body brownish-gray; palpi and antennæ black, a brown spot above each eye and a brownish line in the middle of the front; thorax with three brown stripes, the intermediate double; dorsal segments of the abdomen brownish in the middle, grayish posteriorly and on the sides, sparsely punctured with brownish-black; two larger dots of the same color in the middle of each segment. Legs brown, except the femora, which are reddish-yellow, brown at tip; wings subhyaline, with blackish spots, dots, and cross-bands, a double spot near the root, another spot at the base of the præfurca, an irregular cross-band beginning at the end of the auxiliary vein and ending at the hind margin in the spurious cell; a second interrupted cross-band begins in the region of the stigma and ends on the hind margin in the two last posterior cells; numerous blackish dots in the cells and at the end of the longitudinal veins. Length of the body about 10^{mm}.

Hab.—California (San Mateo Creek, near San Francisco, April 9, 1876). A single male.

Section IX.—*Tipulina*.

Are abundantly represented in California, both in species and specimens. *Pachyrrhinæ*, however, so numerous in the Atlantic States, seem
to be rare.

It would do but little good to describe the numerous species of my
collection until the *Tipulæ* of the Atlantic States are better known. I
confine myself, therefore, to a small number of remarkable and easily
recognizable species, besides reviewing the Californian species described
by Mr. Loew. I add the description of a remarkable *Pachyrrhina* from
the Rocky Mountains.

TIPULA PRÆCISA Loew, Centur., x, 2; ♂ ♀.—A common species in
Marin and Sonoma Counties in April and May. I have two males from
Brooklyn, Alameda County, Cal. (Wm. Holden, M. C. Z.), with abor-
tive wings, which are hardly twice as long as the halteres; in all the
other characters, as well as in the structure of the hypopygium, they
agree with the males of *T. præcisa*. I suppose this is a case of dimorphism. Mr. H. Edwards mentioned to me a subapterous *Tipula*, which
he found in company with the apterous *Bittacus;* it is probably this
very species. I found the *Bittacus* abundantly near Petaluma, and ascertain now, by the dates, that the *Tipula* occurred to me on the same
day, and probably in the same locality; but I did not find any of the
subapterous specimens. I recommend to collectors the verification of
the dimorphism which I suspect.

TIPULA PUBERA Loew, Centur., v, 16.—Common in May in Marin
County, California (San Rafael, May 26, 27; also received from Mr.
H. Edwards and Mr. J. Behrens).

TIPULA FALLAX Loew, Centur., iv, 10.—California. I do not have it.

TIPULA BEATULA n. sp., ♂ ♀.—Wings with longitudinal gray shades
in the center of all the cells of the apical portion; thorax behind the
suture, including scutellum and metathorax, dark brown, shining, yellow
in the middle. Length of male 11–12^{mm}; of the female, including the
ovipositor, 15–16^{mm}.

Antennæ but little longer than the head, brown; scapus brownish-yellow; joints of the flagellum 1–7 subcylindrical, elongate, inconspicuously
verticillate; tip of the flagellum suddenly attenuate, with two or three
joints indistinct in dry specimens; thus the antennæ seems to be only
11- or 12-jointed. Front and vertex grayish-yellow, darker in the middle; rostrum brown above and below; palpi brown. Thorax yellowish-
gray, with three brown stripes and two shorter stripes on each side;
behind the suture, and in front of the scutellum, the mesonotum is dark
brown or black, shining, with a yellow line in the middle; scutellum
yellow, shining, with a brown stripe in the middle; metanotum dark
brown, shining in the middle, pollinose on the sides, and with a yellow
stripe. Knob of halteres brown, with a whitish spot at the tip. Abdomen brownish-yellow; hind margins of the segments and an inter-

rupted dorsal stripe brown. On the venter, the anterior half of the second segment is dark brown, shining; a narrow brown shining cross-band at the base of the three following segments (broader in the female). Terminal club of the abdomen in the male of moderate size, brownish-yellow, with two very small, yellowish, projecting, coriaceous organs above; in the female, the three last joints of the abdomen are shining; the upper valves are almost rounded at tip; the lower ones shorter and pointed. Legs yellowish-brown; tips of femora dark brown. Wings grayish, a gray cloud at the distal end of the first basal cell, other clouds about the middle of the anal cell and at the tip of the seventh vein; longitudinal faint gray shades along the middle of all the cells in the apical portion of the wing, including the axillary; stigma oval, blackish.

Hab.—Marin and Sonoma Counties, in April and May, rather common. Will be easily distinguished by the peculiar coloring of the posterior part of the thorax.

TIPULA SPERNAX n. sp., ♀.—Thorax shining black, smooth, polished; metanotum with a stripe of gray pollen in the middle; wings grayish-hyaline, unicolorous. Length, including ovipositor, about 16^{mm}.

Female.—Head, including rostrum, black, shining; a gray pollen on the front and on the upper part of the rostrum; palpi black; antennæ black, but little longer than the head; joints of the flagellum hardly incrassate at base, with moderate verticils; first joint grayish-pollinose. Thoracic dorsum shining, black, smooth; pleuræ and coxæ grayish-pollinose, a yellowish stripe between the root of the wings and the collare; the space between the root of the wings and the scutellum and metanotum is likewise yellowish; metanotum black, with a stripe of gray pollen in the middle, its sides yellowish; halteres yellowish-brown; abdomen black, shining, slightly grayish-pruinose toward the tip; each segment with a narrow yellow border posteriorly and yellow lateral margins; legs dark brown; femora, except the tip, reddish. Valves of the ovipositor yellowish-ferruginous, the upper ones very narrow, almost linear; tip slightly incrassate, brown. Wings uniformly grayish-hyaline; stigma brown, a whitish spot at its proximal end; veins brown, except those of the root, which is yellow, with a dark brown spot immediately above the insertion.

Hab.—Webber Lake, Sierra Nevada, July 26. A single female.

A very remarkable species, allied to *Pachyrrhina.* Without having the male, it is difficult to ascertain its relationship.

PACHYRRHINA ALTISSIMA n. sp., ♂ ♀.—Altogether black, including legs and halteres; wings brownish, with a small brown stigma. Length, ♂ 12-13^{mm}; ♀ 15-16^{mm}.

Male.—Antennæ about as long as head and thorax together, nodose, verticillate; thorax somewhat shining, slightly grayish-pollinose; abdomen moderately shining, with velvet-black, opaque cross-bands on the segments 2-4; second posterior cell sessile or subsessile; appendages of hypopygium dark brown or black.

Female.—Antennæ but little longer than the head; joints of flagellum slightly incrassate at base; the whole body more opaque than that of the male; hence the velvety cross-bands on abdomen less apparent; ovipositor ferruginous-brown (valves divaricate in the dry specimens).

Hab.—Rocky Mountains, at great altitudes; male specimens were found July 15, 1875, on Taos Peak, New Mexico, above timber-line (W. L. Carpenter); females on Pike's Peak, July 14, 1875, at 13,000 feet altitude (A. S. Packard). Three males and as many females.

PACHYRRHINA FERRUGINEA Fabr.—A very common eastern species. I have a male and a female from California (San Mateo) and two males from Denver, Colo. (P. R. Uhler), which have no black or brown triangles on the abdomen. It seems, nevertheless, to be the same species.

HOLORUSIA RUBIGINOSA Loew, Centur., iv, 1.—Not rare in Marin and Sonoma Counties in May.

CTENOPHORA ANGUSTIPENNIS Loew, Centur., x, 3.—Not rare among the red-woods in the Coast Range. The larva probably lives in the stumps — *N c* of *Sequoia sempervirens*. (Lagunitas Creek, Marin County, April 15; also numerous specimens from Mr. H. Edwards.)

Family BIBIONIDÆ.

BIBIO HIRTUS Loew, Centur., iv, 2; ♂ ♀.—Marin and Sonoma Counties, California, in April, common; Yosemite Valley, June 1–12. Mr. Loew's description implies that this species is very variable, and indeed I have male specimens, from the same localities, which look like a different species; they are almost destitute of whitish hairs, showing only a few of them on the sides of the abdomen; the size of the specimens is also variable.

BIBIO NERVOSUS Loew, Centur., iv, 4; ♀.—California. I did not come across this species.

BIBIO sp., ♂ ♀.—(Saucelito, April 2; San Geronimo, Marin County, April 19.) A red-legged species, like the preceding, but different.

DILOPHUS sp., ♂ ♀.—Marin and Sonoma, in April, common.

Family XYLOPHAGIDÆ.

RACHICERUS HONESTUS n. sp.—Antennæ 22–23-jointed, moniliform, subpectinate; thorax brownish-yellow, with two brown stripes; abdomen dark brown; wings tinged with brownish. Length 5.5^{mm}.

Male.—Head and antennæ black; palpi yellow; front above the antennæ with a silvery cross-band; antennæ 22-jointed, if the last joint, which is double, is counted for one; they are moniliform and subpectinate on both sides, the projections on the lower side being a little longer than those on the upper. Thorax brownish-yellow, with two broad brown stripes, which begin a little behind the humerus and do not touch the anterior margin; pleuræ brown. Abdomen uniformly dark brown, with a delicate grayish-yellow pubescence. Legs, including the coxæ, yellow. Knob of the halteres brownish on the under side. Wings with a uniform slightly brownish tinge.

Hab.—San Rafael, Cal., May 29. One specimen.

The antennæ of *R. honestus* (♂) seem to be like those of *R. ruficollis* (♂) Hal., of which I regret not to have a specimen for comparison ; only instead of having 34–35 joints, they have 22–23. The eyes are notched on the inner side, as described in Walker, List, etc., v, p. 104. The coloring of the body of *R. honestus* is exactly like that of a specimen from Illinois, which I have, and believe to be a mere variety of *R. obscuripennis* Lw.

The North American *Rachiceri* at present known may be grouped thus : —

Antennæ not pectinate:
> *ruficollis* Hal.—Atlantic States.
> *honestus* n. sp.—California.

Antennæ pectinate in the male, not pectinate in the female:
> *varipes* Loew.—Cuba.

Antennæ pectinate in both sexes :
> *obscuripennis* Loew.—Illinois.
> *nigripalpus* Loew.—Mexico.

Family STRATIOMYIDÆ.

CHLOROMYIA VIRIDIS (syn. *Sargus viridis* Say).—A number of specimens taken between San Rafael and Saucelito, Marin County, Cal., April 2, agree with the specimens from the Atlantic States. The same occurs in Colorado (P. R. Uhler).

The name *Chrysomyia* Macq., 1834, cannot be maintained for the genus to which *Sargus viridis* belongs, as there is an earlier *Chrysomyia* Rob. Desv., 1830, among the *Muscidæ*. I reinstate, therefore, the name *Chloromyia*, introduced by Mr. Dnucan (Magaz. Zool. and Bot., 1837).

OXYCERA CROTCHI n. sp., ♀ .—Abdomen with three lateral yellow spots on each side and an apical triangular one, all connected by a narrow yellow margin ; femora black, with yellow tip; tibiæ and tarsi yellow. Length 8^{mm}.

Female.—Face and front yellow, with a broad black stripe in the middle ; posterior orbits yellow ; vertex, cheeks under the eyes, and occiput black. Antennæ: basal joints black (the rest wanting). Thorax black, opaque; a yellow stripe from the humerus to the antescutellar callus is interrupted a little beyond the middle, a pair of narrower yellow stripes on the dorsum slightly expanded in front and not reaching beyond the transverse suture ; scutellum yellow, the base black; pleuræ with a large yellow spot in front of the wings, and a smaller oblong one under it; the black opaque abdomen has a subtriangular yellow spot on each side of the second segment, a larger, semi-elliptical spot on each side of the third segment, a somewhat similar, but smaller, pair of spots on the fourth segment, a large triangular spot on the last segment; all these spots are connected by the narrow, yellow, abdominal margin; ventral segments yellow in the middle, brownish-black on the sides. Femora black, with yellow tips; tibiæ and tarsi yellow; joints 3 and 4 of front tarsi darker. Wings tinged with yellowish anteriorly, with grayish posteriorly; stouter veins and stigma reddish yellow.

Hab.—California (G. R. Crotch). A single specimen.

STRATIOMYIA BARBATA Loew, Centur., vi, 9.

STRATIOMYIA INSIGNIS Loew, Centur., x, 7.

STRATIOMYIA MACULOSA Loew, Centur., vii, 19.

STRATIOMYIA MELASTOMA Loew, Centur., vi, 10.

ODONTOMYIA ARCUATA Loew, Centur., x, 4.

ODONTOMYIA MEGACEPHALA Loew, Centur., vi, 20.

CLITELLARIA LATA Loew, Centur., x, 9.

Of these species, I possess *S. maculosa*, ♀ (San Rafael, in April and May, not rare); *S. insignis*, ♂, which, the difference of the coloring of the face notwithstanding, I hold to be the male of *maculosa;* and *S. melastoma* (Summit Station, Sierra Nevada, July 4, H. Edwards).

I have furthermore three *Stratiomyiæ* and one *Odontomyia* from the Sierra Nevada (Webber Lake, in July), which I abstain from describing.

CLITELLARIA RUSTICA n. sp., ♂ ♀.—(Altogether black, with a short grayish-yellow pubescence; knob of halteres white; wings tinged with dark gray. Length, male, about 6^{mm}; female, 7–8^{mm}.

Black, but little shining, rather evenly clothed with a grayish-yellow, short, semi-recumbent pubescence; the disk of the abdomen above glabrous. Antennæ black; eyes densely pubescent; legs black, beset with a dense, short, grayish pubescence; upper side of tarsi glabrous, deep black. Spines of the scutellum brownish. Wings with a distinct dark gray shade; stigma pale brown; anal cell closed some distance from the margin.

Hab.—The Geysers, Sonoma County, California, May 5–7, common; also in San Geronimo, Marin County, April 19. A male from Webber Lake, Sierra County, July 22, is only 4.5^{mm} long. Three males and six females.

NEMOTELUS sp.—Specimens from the environs of San Francisco (W. Holden) agree with the description of *N. canadensis* Loew, but the edge of the abdomen does not show any trace of yellow.

Family TABANIDÆ.*

California (and with it, probably, the whole western region) is not very rich in *Tabanidæ*, especially when compared to the Atlantic States. Both species and specimens seem to be much less numerous. From the valleys of the Coast Range, in the environs of San Francisco, I possess only four *Tabani*, one *Chrysops*, one *Pangonia*. A *Silvius* is common,

---

* About the *Tabanidæ* from the Atlantic States, compare my Prodrome of the Tabanidæ of the United States in the Memoirs of the Boston Soc. N. H., vol. ii.

especially in the foot-hills of the Sierra Nevada; the same species occurs in Colorado. From the high Sierra I brought three *Tabani* and two *Chrysops*. Besides these species, I describe a *Chrysops* common in the Colorado region and Utah. For comparison's sake, I will state that at least sixty species of *Tabanus* and twenty-five *Chrysops* occur east of the Mississippi. None of these species has as yet been found in California. Dr. Philippi's essay on the fauna of Chili enumerates twenty-two *Pangoniæ* (including *Mycteromyia*), thirty-five *Tabani*, but only two *Chrysops*. The occurrence of *Silvius* in the western region is remarkable, as it is one of the numerous points of analogy between its fauna and that of Europe. A *Silvius isabellinus*, said to be from the Atlantic States, has been described by Wiedemann, but it has never been found since, and if it exists it must be a very rare species. In Central Europe, a single *Silvius* occurs; another one is an alpine species of apparently local occurrence, described by Mr. Loew.

Chrysops fulvaster, from Colorado and Utah, has a fenestrate spot in the discal cell, like several European species, mostly belonging to the Mediterranean fauna. Not a single one of the twenty-five *Chrysops* from the Atlantic States has this peculiarity.

Chrysops noctifer, from the Sierra Nevada, seems to resemble most closely two species from Lapland, one of which, according to Dr. Loew, also occurs in Sitka.

PANGONIA.

PANGONIA HERA n. sp.—Proboscis short, hardly projecting beyond the palpi; body uniformly blackish-gray; wings grayish-brown; eyes pubescent, first posterior cell open. Length 13–14^{mm}.

Female.—Antennæ black; two basal joints grayish; face and front dull yellowish-gray, the former with pale yellowish-white hairs, the latter (at least in my specimen) with a black, shining, denuded space in the middle; the proboscis hardly reaches beyond the short and stout gray palpi; lips large, developed; eyes pubescent, in life unicolorous, green; ocelli distinct. The black ground-color of the thorax and of the abdomen is partly concealed under a dust-colored, dingy-gray pollen; both are clothed with pale yellowish-white hairs. Wings tinged with a dingy brownish-gray; some lighter shades are visible in an oblique light, especially along the rounded part of the hind border and in the second basal cell; first posterior cell open; a long stump of a vein at the base of the fork of the third vein.

Hab.—San Francisco, Cal., caught in the street twice, by Mr. Henry Edwards, in July. I have a single female.

PANGONIA INCISA (syn. *Pangonia incisuralis* Say, Journ. Acad. Phila., iii, p. 31; Amer. Ent., plate xxxiv; *Pangonia incisa* Wiedemann, Auss. Zw., i, p. 90).—Colorado Springs, Colo., in August (P. R. Uhler); Arkansas (Say). The descriptions are easily recognizable.

SILVIUS.

SILVIUS GIGANTULUS (syn. *Chrysops gigantulus* Loew, Centur., x, 12; *Silvius trifolium* Osten Sacken, Prodr. of Tabanidæ, i, 395).—My description was drawn from alcoholic specimens, and requires some emendations.

The three-fold spot at the base of the abdomen is sometimes quite distinct, in other cases very faint. A longitudinal stripe in the middle of the abdomen, formed by an appressed golden-yellow pubescence, is very distinct in most, but not in all specimens. The black spot on the venter is often wanting. Front tarsi black, faintly reddish at base. One of the specimens (Colorado) has no denuded spots on the face; the ocellar area likewise is not denuded; the legs are altogether reddish, except the infuscated ends of the tarsi.

Hab.—California, especially in the Sierra Nevada and along its foothills (Mariposa County, June 3, not rare; Yosemite Valley, June 13; also about Webber Lake, Sierra County, July 26); Washington Territory; Vancouver Island (J. R. Crotch); Southern Colorado (W. L. Carpenter).

The living insect has bright yellowish-green eyes, with numerous irregularly scattered black dots; this agrees with Meigen's and Loew's description of the eyes of the other species of the genus.

In my former description, line 2, for *regular* spots, read *angular*.

That Mr. Loew took this species for a *Chrysops* is, I suppose, nothing but an oversight on his part.

TABANUS.

Of the seven species of *Tabanus* from California which I possess, three belong to the high Sierra, and four were found in the valleys of the Coast Range. Four of these species belong to the subgenus *Therioplectes*, one is apparently an *Atylotus*, and the remaining two are true *Tabani*.

All descriptions of species of the difficult group of *Therioplectes* must be necessarily imperfect as long as they are based on female specimens only. I have done my best to render them clear, and have taken note of the coloration of the eyes of the living specimens. As far as my knowledge goes at present, these two subgenera may be defined as follows:—

Therioplectes.—Eyes pubescent, with three or four purple cross-bands and intervening green intervals; ocelligerous tubercle more or less distinct; head of the male not differing much in size and shape from that of the female; the difference in size between the large and small facets on the eyes of the male is but very moderate, and the dividing line between them indistinct. (Species Nos. 1–4.)

Atylotus.—Eyes pubescent, with a single narrow purple cross-band in the middle, or unicolorous (?); no vestige of an ocellar tubercle; frontal callosity either entirely wanting or imperfect (in the Californian *T. insuetus* it is narrow and does not reach the eyes on either side, which is not the case with an ordinary *Tabanus*). Head rather large, very convex anteriorly and concave posteriorly. In the male, the difference in size between the large and small facets is considerable, the line of division between them distinct; palpi (♀) stout at base; third joint of the antennæ rather broad, with a comparatively short, stout, annulate portion; upper branch of the third vein knee-shaped at base, with a tendency to emit a stump of a vein; first posterior cell broadly open. (Species No. 5 provisionally placed here until the male is found).

Tabanus, in the broader sense, species Nos. 6 and 7.

1. TABANUS PROCYON n. sp.—Is a *Therioplectes*; eyes densely pubescent, even in the female; in life, they have four green stripes on dark purple ground, the upper stripe not sharply limited on the upper side.

Female.—Altogether black; subcallus black, shining, rather swollen; third antennal joint rather narrow. Length 13ᵐᵐ.

Front broad, opaque, slightly grayish, clothed with black pile; callosity large, broader than long, convex; subcallus denuded, shining, somewhat swollen; cheeks black, shining, with long brownish hair; antennæ black, third joint long and narrow, its upper corner hardly projecting; ocellar tubercle distinct; palpi black, stout at base; thorax black, with long black pile, especially on the pleuræ; abdomen black, moderately shining; the two last segments with some whitish hairs; legs black. Wings:—costal cell tinged with brown; bifurcation of the third vein, cross-vein at the base of second posterior cell, also the central cross-veins, more or less distinctly clouded with brownish.

Hab.—Marin and Sonoma Counties, California, April, May. Two females.

Is somewhat like the European *T. micans,* but can be at once distinguished by the deep black color of the pile on pleuræ and chest.

2. TABANUS SONOMENSIS n. sp.—A *Therioplectes* of the group of *T. epistates* (syn. *T. socius* of my Prodrome) and of the European *T. tropicus.*

Female.—Grayish-black; sides of the abdomen red; abdominal incisures fringed with yellowish hairs, which also form faint triangles in the middle of the segments; wings with a small but distinct brown cloud on the fork of the third vein. Length 13–15ᵐᵐ.

Antennæ black, sometimes fainty reddish on the second joint and at the base of the third; the latter is not broad, its upper angle either moderately or very little projecting. Front moderately broad, clothed with brownish-gray pollen, and beset with black, erect pile; callosity large, square, sometimes with a linear prolongation above; subcallus not denuded; cheeks whitish, with yellowish-white hair; palpi stout at base, yellowish-white, beset with short black hairs; eyes pubescent; ocellar tubercle distinct. Thorax grayish-black, the usual gray lines very faint; antealar tubercle reddish; pleuræ thickly clothed with hairs of a dingy gray. Sides of the abdominal segments 1 to 4 rufous, thus leaving between them a black stripe, expanding anteriorly; the remaining segments black;

hind margins of all the segments fringed with yellowish hairs, which also form faint triangles in the middle of the segments; in very well preserved specimens, faint lateral spots, formed by yellow hairs, on segments 2 and 3, are perceptible. Venter reddish, blackish at the base and toward the end. Femora blackish-gray, beset with grayish pile; front tibiæ red at base, their distal half and tarsi black; four posterior tibiæ and first joint of tarsi red, beset with black pile, which, on the hind tibiæ, forms a distinct fringe. Wings with a faint grayish tinge; costal cell and stigma yellowish-brown; a faint brownish cloud across the central cross-veins; a small cloud on the fork of the third vein; the latter often appendiculate.

Hab.—Marin and Sonoma Counties, California, April 27 to May 9, common. Eleven females.

Not unlike my *T. epistates*, but easily distinguished by the black antennæ, darker thorax, distinct cloud on the fork of the third vein, more conspicuous fringes of yellow hair on abdominal segments, etc.

3. TABANUS PHÆNOPS n. sp.—A *Therioplectes* of the same group with *T. sonomensis.*

Female—Grayish-black; sides of the abdomen red; wings hyaline, no distinct brown cloud on the bifurcation of the third vein; antennæ black. Length 13–14^{mm}.

Front gray, a little converging; ocellar tubercle distinct; callosity nearly square, with a spindle-shaped prolongation above; antennæ black; third joint rather narrow, its upper angle very little projecting; thorax grayish-black, with the usual lines very faintly marked; the antealar callosity variable, reddish or dark. The black stripe inclosed between the reddish sides of the abdomen is generally rather broad, and somewhat expanded at the posterior margins of segments 2 and 3, so as to appear jagged; the red on the sides of segments 2, 3, and 4 is clothed with a scarce and very minute golden-yellow pubescence, in the shape of faint, oblique spots; it also forms a fringe on the incisures.

T. phænops is very like *T. sonomensis*, but it is usually a little smaller, the front is narrower, the bifurcation of the third vein is not clouded; in most, but not in all, specimens, the red on the sides of the abdomen is less extended, leaving a broader black stripe in the middle, which is expanded at the abdominal incisures, and therefore appears jagged. In shape, the abdomen is more elongated, with more parallel sides. In life, this species is easily distinguished by the color of its eyes, which are of a very bright green, with comparatively narrow purple cross-bands, much narrower than the green intervals between them; no purple in the upper and lower corners of the eye (at least, in the specimens observed).

Hab.—Webber Lake, Sierra County, California, July 27. Four females. Two specimens from Fort Bridger, Wyo., August 4, seem also to belong here.

4. TABANUS RHOMBICUS (*Tabanus rhombicus* Osten Sacken, Prodrome, etc., ii, 472).—A number of specimens from Webber Lake, Sierra Nevada, July 21–27, closely resemble my *Tabanus* (*Therioplectes*) *rhombicus* from the Colorado Mountains. These specimens show two or even three distinct forms, which I will characterize, in order to draw the attention of collectors to them. I have five specimens of each of these groups, all females.

1. The lateral triangles on the second and third segments of the abdomen are rectangular,—that is, their inner side is perpendicular to the hind margin of the segment, or nearly so ; the prolongation of the outer angle toward the lateral margin forms a broad border on the hind margin ; the intermediate triangles are well defined, equilateral, their apex not prolonged in a line reaching to the next segment ; antealar callosity black ; subcostal cell distinctly tinged with brownish ; subcallus denuded ; none of the specimens has a stump of a vein on the fork of the third vein. Length 12–13.5^{mm}.

2. The lateral triangles on the abdominal segments from 2 to 5 are oblique,—in fact, more streaks than triangles ; their prolongation toward the lateral margin is a narrow whitish border of the hind margin ; the intermediate triangles (rubbed off in most specimens) show a tendency to a linear prolongation of the apex toward the next segment ; antealar callosity faintly reddish ; subcostal cell nearly hyaline ; subcallus either not or only partially denuded ; stump of a vein present in most, not in all, specimens. Length 14–15^{mm}.

Intermediate between these two groups of specimens, there is a third, which combines the abdominal markings and the faintly reddish antealar callosity of the second group, with the infuscated costal cell, the perfectly denuded subcallus, the absence of the stump, and the smaller size of the first. My specimens of this group have the sides of the abdomen very distinctly reddish.

All the specimens were taken indoors, on a window, promiscuously, together with *T. phænops* and *T. insuetus*. I took note of the eyes as having "alternate green and dark purple stripes of about equal breadth". The specimen bearing this label belongs to the intermediate group. Another specimen of the same group and one of the first group are marked as having the green stripes broader.

My specimens from Colorado are nearer to the first form, without reproducing it exactly. The abdomen seems narrower, the abdominal spots less pure whitish. A specimen from Twin Lakes, Colorado, is certainly identical with the second form, and its subcallus is not at all denuded, and it has a distinct stump on the forked vein.

The uncertainty whether the typical *T. rhombicus* from Colorado is the same as the *Tabani* of the first form, the existence of the intermediate group, and the total absence of male specimens, are so many causes why it would be premature to describe the second form as a separate species.

5. TABANUS INSUETUS n. sp.—Belongs apparently to the subgenus *Atylotus*. Eyes pubescent, although in the female specimens the pubescence is often hardly perceptible; in life, pale olive-green, with a single very narrow brown stripe in the middle (distinct even in dry specimens); no vestige of an ocellar tubercle; frontal callosity rather small, variable in size, narrower than the front; third antennal joint rather broad and short, with a short and stout annulate portion; palpi stout at base; first posterior cell broadly open; base of upper branch of third vein knee-shaped, in many specimens with a stump of a vein. All these characters would justify the location of the species in that sub-genus; the discovery of the as yet unknown male will have to decide it.

Female.—Face and front yellowish-gray; cheeks with pale hairs; front with short black hairs; a fringe of such hairs on the upper edge of the occiput. Front broad (in most specimens; much narrower in others); frontal callosity narrower than the front, rather small, and variable in shape; usually another black, shining spot above it. Palpi short, stout at base, pale yellowish or yellowish-white, with black pile. Antennæ, pale brownish-red; annulate portion of third joint sometimes, but not always, black or brown. The black ground-color of the thorax is partly concealed under a gray pollen; vestiges of longitudinal gray lines are visible anteriorly; a pale golden, sometimes whitish, appressed, rather scarce, pubescence, and black, erect pile clothe the dorsum. Pleuræ gray, with pale gray hairs. Abdomen in well-preserved specimens with three rows of yellowish-gray spots, formed by an appressed pubescence; the triangles of the intermediate row large, occupying the whole breadth of the segment; the spots of the lateral rows are oblique, prolonged laterally along the hind border of the segments (well-preserved specimens seem rarely to occur; in the worn specimens, the abdomen appears as grayish-black, somewhat reddish on the sides of the first two segments, and with but vestiges of the appressed yellowish-white pubescence and of the abdominal spots). Venter uniformly yellowish-gray. Feet variable in coloring, pale reddish-yellow, with blackish (seldom pale) femora and tips of tibiæ; tarsi blackish, the two posterior pairs paler at base. Costal cell and stigma more or less tinged with brownish-yellow; upper branch of third vein often, but not always, with a stump of a vein. Length 12–13^{mm}.

Hab.—Webber Lake, Sierra County, July 21. Twelve females.

6. TABANUS ÆGROTUS n. sp.—*Female.*—Altogether brownish-black; wings immaculate; third antennal joint very broad at base. Length 19–20^{mm}.

Front, face, and cheeks clothed with a dense brown pollen, hiding the ground-color; front moderately broad; frontal callosity subobsolete, flat, not shining, prolonged above in an opaque line; cheeks with brown hairs; palpi dark brown; antennæ black, third joint very broad, expanded, and rounded on the under side, and with a projecting upper

3 H B

angle; annulate portion rather abruptly attenuated, as long as the body of the joint. Thorax brownish-black; pleuræ and pectus black, with long black pile. Abdomen black, moderately shining. Legs black. Wings slightly tinged with gray; costal cell and stigma brownish; first posterior cell hardly coarctate.

Hab.—California (H. Edwards). A single female.

In its general appearance and coloring the species is not unlike *T. nigrescens*, from which it is easily distinguished by the absence of brown spots on the wings and other characters. It probably belongs in the same group, especially if the shape of the head of the as yet unknown male and the coloring of the eyes are like the same characters in *T. nigrescens* and *punctifer*.

7. TABANUS PUNCTIFER (*Tabanus punctifer* Osten Sacken, Prodrome, etc., ii, 453, 29).

Hab.—Colorado, Utah; Sonora, California; not rare in the valleys of the Coast Range in June and July.

CHRYSOPS.

The female specimens of the species described below may be tabulated as follows :—

Apex of the wing beyond the cross-band more or less infuscated; first basal cell altogether, or to a considerable extent, infuscated :

    Second basal cell infuscated on its proximal third or beyond :

        Prevailing color of the body black; palpi black.....1. *noctifer*.

        Prevailing color of the body brownish-yellow; palpi reddish ....2. *fulvaster*.

    Second basal cell hyaline :

        The black facial callosities small, not converging anteriorly, separated by a broad, ferruginous interval.......3. *proclivis*.

        The black facial callosities large, converging anteriorly, separated by a narrow ferruginous interval...........4. *surdus*.

1. CHRYSOPS NOCTIFER n. sp.—*Female.*—Cheeks and the converging facial and the large frontal callosities black, shining; between them, the usual yellowish-gray pollen; antennæ black, reddish at base. Thorax black, shining, with vestiges of grayish pollen, forming a faint, broad, geminate stripe anteriorly; scattered whitish pile on the dorsum, more dense above the roots of the wings and on the pleuræ. Abdomen black; sides of segments 1 and 2 red, this color occupying about one-half of the breadth of the dorsum; a faint, appressed, whitish pubescence on the red; similar whitish hairs on the hind margins of the segments (in very well preserved specimens these hairs form very faint triangles on segments 2 and 3). Legs black; four hind tarsi brownish at base. Wings:—costal cell, two-

thirds of first, more than one-half of second basal cell dark brown; the dark brown cross-band does not reach the posterior margin, and does not fill out the third, fourth, and fifth posterior cells; at the proximal end of the fifth posterior cell, near the cross-vein, the brown often but not always shows a small hyaline space; apical spot small, pale grayish-brown, occupying the extreme end of the marginal and first submarginal, and encroaching but very little on the second submarginal cell; the hyaline triangle reaches the first longitudinal vein (if there is any connection between the apical spot and the cross-band, it is a very faint one). Length 9-10^{mm}.

Hab.—Webber Lake, Sierra County, California, July 20-27. Four females.

The eyes of this species in life are like figure 2 of my Prodrome (p. 369), the two central black spots being sometimes disconnected.

This species is closely allied to *C. nigripes* Loew, and still more to *C. lapponicus* Loew, but seems to be different from both, as in those species the apical spot seems to be confluent with the cross-band (if I understand Dr. Loew's expression, "der schmale Spitzenfleck . . . . steht mit der Schwärzung des Vorderrandes in vollständigster Verbindung "— Verh. zool.-bot. Ges., 1858, p. 623). *C. nigripes* occurs not only in Lapland, but, according to Loew, also in Sitka.

2. CHRYSOPS FULVASTER n. sp.—*Female.*—Facial tubercles either entirely red or red mixed with black; frontal tubercle comparatively small, pale red, with a black upper margin more or less extended; palpi pale reddish. Thorax clothed with a yellowish-brown pollen, four stripes, leaving blackish intervals between them. Abdomen:— two basal segments yellowish, the first with a transverse black spot under the scutellum, the second with two black spots, separated by a yellow interval in the middle; the following segments black, with yellow posterior margins, and a more or less distinct yellow longitudinal line in the middle. Legs rufous, more or less black on the joints; front tarsi and the tips of the other tarsi black. Wings somewhat pale brown, almost grayish along the hind margin; a hyaline space occupies the end of the first basal cell, the larger portion of the second basal, the whole anal cell except its distal end, and the proximal half of the anal angle; a crescent-shaped subhyaline space separates the brown cross-band from the more grayish apex of the wing; the inside of the discal cell is paler brown, a spot in the fourth posterior cell subhyaline. Length 6-7^{mm}.

Male.—Black, except the face; palpi black; narrow lateral margins of abdominal segments 1 and 2 and hind margins of the other segments yellow; fulvous pollen in stripes on pleuræ and above the root of the wings; fulvous pile on these stripes. Prevailing color of front legs black; of the two hind pairs reddish. Wings uniformly tinged with black; a crescent-shaped subhyaline space separates the usual cross-band from the apical portion of the wing; another subhyaline elongated spot at the distal end of both basal cells. Length 6^{mm}.

Hab.—Denver, Colo., August 5 (P. R. Uhler); Utah (J. D. Putnam). This seems to be the common species in those regions. The coloring of the body is variable, but the design on the wings will be easily recognized. Five females and one male. It is not possible to identify *C. fulvaster* with *C. quadrivittatus* Say, although it is rather singular that the latter should never have turned up as yet in any of the western collections.

3. CHRYSOPS PROCLIVIS n. sp.—*Female.*—Facial tubercles black on the outside of the dividing furrow only, thus leaving a broad ferruginous interval between them; cheeks black, shining; the intervals between cheeks, facial tubercles, eyes, rontal tubercle, and antennæ are filled out with stripes of pale fulvous pollen. Antennæ black; underside of first joint reddish (sometimes the red is more extended). Thorax black, moderately shining, clothed with yellowish pile; a stripe of gray pollen each side between the scutellum and the humerus is more densely overgrown with yellow pile; the same pile on the pleuræ. Abdomen black; the sides of segments 1 and 2 yellow, leaving an elongated black square in the middle, slightly coarctate on the hind margin of segment 1, and dovetailed on the hind margin of segment 2, by the insertion of a yellow triangle; near the same hind margin, on each side, there is a more or less developed black dot; segment 3 is black, with a yellow hind margin and three more or less distinct longitudinal lines, breaking up the black in four portions; segment 4 black, with a yellowish hind margin, sometimes expanded into a triangle in the middle; the following segments black, with narrow yellow margins; all the yellow portions, including the hind margins, are beset with short yellow hairs. Venter likewise variegated with black and yellow. Front legs black; base of tibiæ reddish; on the posterior pairs, the prevailing color is red, with more or less black on the joints and at the base of the femora. Wings:—costal cell and first basal cell brown, the latter with a small hyaline space at the distal end crossed by a brown line; second basal, anal cell, anal angle, and fifth posterior cell hyaline; the cross-band reaches the hind margin and fills out the fourth posterior cell; apical spot narrow, encroaching but very little on the second submarginal cell; the hyaline triangle enters the marginal cell, but is separated by a brown shade from the costa; distal margin of cross-band slightly protruding toward the base of the second submarginal cell. Length 8–9^{mm}.

Hab.—Marin County, California. Four females.

As usual in species of this kind of coloration, the extent of the yellow on the abdomen is somewhat variable.

4. CHRYSOPS SURDUS n. sp.—*Female.*—Very like *C. proclivis*, but differs in being smaller; the facial callosities are black and shining on both sides of the dividing furrow; being prolonged anteriorly, they coalesce above the mouth; the ferruginous space between them is a narrow stripe, interrupted anteriorly. The thoracic dorsum anteriorly shows two distinct gray longitudinal lines, reaching to about the middle of the thorax; the pile on the pleuræ is of a paler yellow. On abdominal segments 1 and 2, the elongated black square is more distinctly coarctate on the hind margin of the first segment; on segment 3 there is a yellow dividing line in the middle, but the lateral yellow marks in most cases do not exist. The prevailing color of all the legs is black, with only a little reddish at the base of the four posterior tibiæ and tarsi. The design on the wings does not show any important difference. Length 7–8^{mm}.

Hab.—Webber Lake, Sierra County, California, July 21. Four females.

The eyes of this species have the normal coloration (like the figure 1 in my Prodrome).

Family LEPTIDÆ.

As far as the small number of known *Leptidæ* from California enables me to judge, this family exhibits, on the Pacific slope, a more European than Eastern American character.

The striking forms of golden-haired *Chrysopilæ*, the principal feature of the fauna of the Atlantic States, are replaced here by small and insignificant species.

The genus *Triptotricha*, however, hitherto peculiar to North America, seems equally well represented in the Atlantic and Pacific States.

The considerable number of Californian species of *Symphoromyia* and the abundance of specimens are remarkable.

TRIPTOTRICHA LAUTA Loew, Centur., x, 15.—California.

TRIPTOTRICHA DISCOLOR Loew, Berl. Ent. Zeitschr., 1874, p. 379.—California.

I have neither of these species. A specimen which I found near Lake Tahoe, Sierra Nevada, July 19, seems to be different from both.

LEPTIS COSTATA Loew, Centur., ii, 4.—Not rare in Marin and Sonoma Counties, California. The front and hind legs of my five specimens are not as dark as described; but the coloring of the legs seems to be very variable.

LEPTIS INCISA Loew, Centur., x, 16.—The female alone is described; the male has usually much darker femora; the coloring of these, however, is very variable in both sexes. One of my females has a pale reddish scutellum; another has it black at base, reddish toward the tip. Not rare in Marin County in April.

CHRYSOPILA HUMILIS Loew, Berl. Ent. Zeitschr., 1874, p. 379.—*Male.*—"Atra, opaca, tota pilis lutescentibus vestita; tibiæ testaceæ, apicem versus fuscæ; tarsi toti fusci; alæ saturate cinereæ, stigmate fusco. Long. corp. 2 lin.; long. al. 1$\frac{11}{12}$ lin." (About 4.4 and 4.2^{mm}.)

" Black, opaque; antennæ, palpi and knob of halteres of the same color. The erect pubescence of the whole body, and likewise that of the palpi and coxæ pale luteous-yellow, only that of the frontal tubercle a little darker, so that it looks almost black when held against the light. Femora black, at the tip pale luteous-yellowish; their short tomentum of an impure whitish. The luteous-yellowish tibiæ become gradually brown towards the tip, and the feet are tinged with brown, except sometimes at the base of the first joint. Wings with an intense grayish tinge; the stigma dark brown, of a medium breadth and length; the wing-veins blackish-brown.

"*Hab.*—San Francisco (H. Edwards)."

A species of which I found several males near Los Angeles in March differs from this description in having the pubescence of the body golden-yellow, rather than luteous; that of the femora likewise golden-yellow; that of the palpi decidedly black; the stigma is brown, but not dark brown.

A specimen from Webber Lake, California, July 24, has longer and less brownish wings, but a darker stigma; first antennal joint with long bristles, which do not exist in the specimens from Los Angeles; palpi very long; pleuræ grayish; pubescence of the abdomen whitish.

I cannot identify either of these species with the above description.

ATHERIX VARICORNIS Loew, Centur., x, 13.—*Female.*—I do not know this species.

SYMPHOROMYIA sp.—Half a dozen species which I took in Marin and Sonoma Counties in April and May, and about Webber Lake in July, all have the anal cell open, and therefore belong to the genus *Symphoromyia* Frauenfeld (*Ptiolina* Schiner, not Zetterstedt). California seems to be much richer in this group than Europe or the Atlantic States of North America. But as these species resemble each other very closely, and as both sexes often differ in coloring, I deem it more prudent not to attempt to describe them.

The female of one of these species which I observed near Webber Lake stings quite painfully, and draws blood like a *Tabanus.* I am not aware of the fact having ever been noticed before concerning any species of *Leptidæ.*

Family NEMESTRINIDÆ.

Hirmoneura brevirostris Macquart, Dipt. Exot., suppl., i, p. 101, tab. 20, f. 1, from Yucatan, is the only species of this family hitherto recorded as occurring north of the Isthmus of Panama. I describe a species from Texas, of which I have a single specimen, the only Nemestrinid from North America I have ever seen. This scarcity is the more remarkable, as the regions of Central Asia, which, in other respects, show many faunal analogies with the western plains and California, are very rich in species of that family. Many species occur in the countries round the Mediterranean. Dr. Philippi enumerates not less than twenty-one species from Chili (!).

HIRMONEURA CLAUSA n. sp.—Body clothed with pale yellowish-gray hair; antennæ and feet reddish; eyes bare; second submarginal and second posterior cells closed and petiolate at the distal end. Long. corp. 9–10^{mm}.

Face densely covered with pale yellowish hair, through which a short, reddish proboscis is hardly visible; antennæ reddish; front clothed with the same pale yellowish hair; vertex black, with a tuft of black hair; behind it, on the occiput, a tuft of yellow hair. Eyes bare. Thorax clothed with the same pale yellowish hair, especially dense on the pleuræ and pectus; on the dorsum, the black ground-color is visible; the posterior corners, as well as the hind margin of the scutellum, are reddish-brown. Abdomen black, clothed with the same pale yellowish hair. Halteres reddish. Legs brownish-red; femora clothed with pale yellowish, erect pile, especially on their proximal half. Wings hyaline; the veins near the costa reddish-brown; the second submarginal cell is closed, eye-like, long-petiolate at the distal end; the second posterior cell (that is, the cell which is separated from the second submarginal by a single cell, the first posterior, which opens at the apex of the wing) is also closed, with a petiole at the distal end half as long as the petiole of the second submarginal; the third posterior cell is closed (as usual in this genus).

Hab.—Dallas, Texas (Boll). A single specimen, apparently a female.

The venation of this species is like that of *H. brevirostris* Macquart (Dipt. Exot., suppl., i, tab. 20, f. 1), except that the second posterior cell is closed, and the petiole of the second submarginal is longer than represented on the figure.

Family BOMBYLIDÆ.

The *Bombylidæ* are perhaps the most characteristic and one of the most abundantly represented families of *Diptera* in the western region, including California. Nevertheless, the results obtained by me in working up this family are not at all in proportion to the number of species collected. I have been very much hampered, on the one hand, by the unsatisfactory condition of the systematic distribution of the *Bombylidæ* in general; on the other, by the insufficiency of my eastern material and the impossibility of identifying the large number of existing descriptions of eastern species.

For fear of increasing the difficulties of the future student, I have confined myself to the description only of the more striking forms; and, at the same time, in order to facilitate his task, I have taken advantage of this opportunity for reviewing all that has been hitherto done for North American *Bombylidæ*. A list of all the described species from

North America north of Mexico, distributed as far as possible among the genera where they belong, will be found on the following pages.*

An analytical table of all the genera hitherto found in the United States is also given.

From the very circumstance that the *Bombylidæ* are one of the most numerously represented families of *Diptera* in the Western Territories, it follows that it would be premature now to attempt any generalization about their geographical distribution. The following remarks, based upon the existing data, are therefore only provisional.

Among the group of *Anthracina*, the genera *Anthrax*, *Exoprosopa*, and *Argyramœba* are abundantly represented both in the Atlantic and Pacific States, but probably more so in the latter. The new genus *Dipalta*, with a single species, occurs in Colorado and in California as well as in Georgia.

The North American species of the group *Lomatina*, which I have seen, have the general appearance of *Anthrax*, but, at the same time, a very short præfurca, with the small cross-vein far beyond its end, and the eyes contiguous in the male. They differ from the *Bombylina* in the globular shape of the head, the very large size of the frontal triangle of the male, and often in the *Anthrax*-like antennæ, more or less distant from each other at the base. The genus *Oncodocera*, with *O. leucoprocta* from the Atlantic States, belongs here. I have introduced the new genus *Triodites*, with one species from California and Utah. I possess a species from Colorado, which will require the formation of a new genus; I do not describe it at present. *Anisotamia eximia* Macq. (= *Anthrax valida* Wied.) from Mexico is related to *Oncodocera*. From Mr. Loew's statements about *Aphœbantus* and *Leptochilus*, both new genera, with a single species from Texas, I judge that they likewise belong to this group. The *Stygia elongata* Say (*Lomatia elongata* Wied.) is evidently not a *Lomatia*, and perhaps not a Bombylid at all. I have never seen it.

The *Toxophorina* are represented as yet only by one *Toxophora* from California and by several from the Atlantic States. A single *Systropus* occurs in the Atlantic States.

The *Bombylina* are represented by the genera *Bombylius*, *Systœchus*, and *Sparnopolius*. From *Systœchus* I have separated the genus *Anastœchus* which also occurs in Europe. *Pantarbes* nov. gen., with a single species from California, is not unlike *Mulio*. *Lordotus* Lw., with one species, occurs in Colorado, Wyoming, and Texas, as well as in California. *Comastes* nov. gen., with one species from Texas, is a very original and interesting form. *Ploas* is represented by seven species from California,

---

* In identifying species from the United States, the descriptions of species from the West Indies and Mexico must not be quite neglected, as some of these species may have a wide northerly range. Lists of these species will be found in my Catalogue of the Described Diptera of North America, Smithsonian Institution, 1858. The species published since will be found in Loew's "Centuries", in Jaennicke's "Exotische Diptern", and in Bellardi's "Saggio".

one from New Mexico, and one from the Southern States. *Geron* occurs everywhere. Two *Phthiriæ* are known from the Atlantic States, one more from Colorado, and three from California.

The most interesting addition to the North American fauna in this family is *Epibates*, a new genus, the male sex in which is distinguished by a muricate surface of the thoracic dorsum. I have not less than seven species in it, four from the Pacific coast, two from the Atlantic States, and one uncertain.

As a general result, I will state that the large genera of this family occurring in the United States are universal or nearly universal genera.

The genera peculiar to the fauna, with the exception of *Epibates*, are all as yet monotypical. The genera which do not belong to either of these two categories are:—*Ploas*, which, besides North America, occurs, as far as I know, only in the fauna of the Mediterranean and Central Asia. It is singular that it has not been recorded from South America. *Systropus* counts several species in Mexico and South America, also at the Cape, and in Australia. *Toxophora* occurs in Algiers, Syria, the Cape, Brazil, and Java. *Phthiria* is found in the Mediterranean region and in Central Asia, at the Cape, also in Brazil and Chili.

A fact worth noticing is the common occurrence of some species of *Bombylidæ* in both hemispheres, or, if the specific identity is contested, at least the great resemblance between some species in Europe and America.

The European *Bombylius major* seems to be the same as the most common species in California. *B. fratellus*, from the Atlantic States, is very little different from it. *Systœchus vulgaris* and *Anastœchus barbatus* are remarkably like the European species of the same genera. *Anthrax dorcadion* n. sp. (= the true *A. capucina* F.) is the same, or nearly the same, as the species known as *A. capucina* in Europe.

Of all families of *Diptera orthorhapha*, hardly any have been so imperfectly studied in their organization as the *Bombylidæ*. By gradual additions, the number of genera has reached very near seventy, and nevertheless the discrimination of the essential characters on which to base a classification may be said not to have been even begun. Dr. Schiner (Novara, p. 115) proposed a subdivision of the family in four groups,—the *Anthracina*, *Lomatina*, *Toxophorina*, and *Bombylina*. But, as he did not characterize these groups, this subdivision can have but very little value. It would seem self-evident that any attempt at a subdivision must be preceded by a thorough study of the outward organization of these insects; nevertheless, this has never been done yet. The thick fur, the hairs and scales, which cover the whole body, or certain parts of it, render such a study difficult, unless that covering is removed; and many an important character may have been overlooked, owing to the neglect of undergoing that trouble. As an instance of such an oversight, I will mention the remarkable epimeral hooks which exist in most of the genera of the *Anthracina* above the root of the

wiugs; as far as I am aware, their very existence has never been mentioned anywhere.

Under such circumstances, and especially in the absence of collections
containing some of the foreign generic forms, the task of establishing
the indispensable new genera becomes a very difficult one. I have
spared no trouble in reading the descriptions of the existing general
and hope, as far as lay in my power, to have avoided redescribing old
genera under a new name.

The following table contains all the genera of *Bombylidæ* hitherto
found in North America north of Mexico. The genera hitherto recorded
as occurring in Mexico, and not found yet in the United States, are:—

Adelidea Macq., Dipt. Exot., ii, 1, p. 84, for *A. flava*, Jaennicke, from
Mexico. According to Schiner (Novara), this genus is the same as .*So-
barus* Loew, Beitr., iii.

Anisotamia Macq., Dipt. Exot., ii, 1, p. 81, for *A. eximia* (= *Anthrax
valida* Wied.), is closely related to *Oncodocera*. Whether it is a true
a genus established by Macquart for certain African species, remains
Anisotamia, to be proved.

Pœcilognathus Jaennicke is simply a *Phthiria*.

In using this table, it must be borne in mind that I have not seen
Aphoebantus Lw. and *Leptochilus* Lw., and have placed them according
to the data of the descriptions; and that I do not know the male sex of
Comastes.

Analytical table of the genera of BOMBYLIDÆ, occurring in North America north of Mexico.

1 (10). The bifurcation of the second and third veins takes place opposite, or nearly opposite, the small cross-vein; the second vein
forms a knee at its origin from the præfurca; the third vein
is in a straight line with the præfurca:

2 (5). Three submarginal cells, the anterior branch of the third vein
being connected with the second vein by a recurrent cross-
vein:

3 (4). Antennæ with a more or less long style at the end of the third
joint .... *Exoprosopa* Macq.

4 (3). Antennæ without any distinct style at the end, *Dipalta* nov. gen.

5 (2). Two submarginal cells:

6 (7). Third antennal joint with a distinct pencil of hairs at the tip;
pulvilli distinct .... *Argyramœba* Schiner.

7 (6). Third antennal joint without pencil of hairs at the tip:

8 (9). Pulvilli distinct.................... *Hemipenthes* Loew, Centur.

9 (8). No distinct pulvilli .... *Anthrax* Scopoli.

10 (1). The bifurcation of the second and third veins takes place some
distance before the small cross-vein, at an acute angle; the
second vein does not form a knee at its origin from the
præfurca:

11 (18). Body *Anthrax*-like; frontal triangle in the male unusually large; frontal space in the female of a corresponding size :

12 (13). Antennæ approximated at base; third antennal joint gradually attenuated.............................*Oncodocera* Macq.

13 (12). Antennæ remote at base; third joint subglobular at base, suddenly contracted, and then linear, styliform :

14 (17). Pulvilli distinct :

15 (16). Second submarginal cell appendiculate,
Aphœbantus Loew, Cent.

16 (15). Second submarginal cell not appendiculate, *Triodites* nov. gen.

17 (14). Pulvilli none.........................*Leptochilus* Loew, Cent.

18 (11). Body not *Anthrax*-like; frontal triangle in the male small :

19 (22). Body (antennæ, thorax, abdomen) clothed with more scales than hairs, gibbose, the abdomen hanging down; antennæ long, first joint unusually long :

20 (21). Four posterior cells ....*Lepidophora* Westw.

21 (20). Three posterior cells ....*Toxophora* Meig.

22 (19). Body clothed with hairs, or else nearly glabrous :

23 (44). Four posterior cells :

24 (33). First posterior cell closed :

25 (32). Two submarginal cells :

26 (29). First basal cell longer than the second :

27 (28). Front and epistoma in the profile form a gently inclined plane; the latter with long and dense, bushy pile; head narrower than the body..............................*Bombylius* Lin.

28 (29). Front and epistoma in the profile nearly vertical, without bushy pile; head large, as broad as the body; thorax large, more bulky than the abdomen..................*Comastes* n. gen.

29 (26). Both basal cells of equal length :

30 (31). Under side of the head moderately pilose, and hence its different parts (including the base of the antennæ, the oral edge, etc.) easily perceptible ....*Systœchus* Loew, Beitr.

31 (30). Under side of the head densely pilose, the root of the antennæ, epistoma, mouth, etc., being completely hidden,
Anastœchus n. gen.

32 (25). Three submarginal cells.....................*Pantarbes* n. gen.

33 (24). First posterior cell open :

34 (41). Two submarginal cells :

35 (36). Both basal cells of equal length.....*Sparnopolius* Loew, Beitr.

36 (35). First basal cell longer than the second :

37 (38). Third antennal joint not truncate at the tip ....*Epibates* n. gen.

38 (37). Third antennal joint flattened, truncate at the tip :

39 (40). Proboscis short ....*Paracosmus* (*Allocotus* Lw.).

40 (39). Proboscis very long............................*Phthiria* Meig.

41 (34). Three submarginal cells :

42 (43). Proboscis long; abdomen convex ....*Lordotus* Loew, Cent.

43 (42). Proboscis short; abdomen flattened..............*Ploas* Latr.
44 (23). Three posterior cells; anal cell closed :
45 (46). Proboscis much longer than the antennæ; small, *Bombylius*-like, pubescent insects............ ....*Geron* Meig.ı
46 (45). Proboscis shorter than the antennæ; long, *Sphex*-like, almost glabrous insects; abdomen with the four basal joints very narrow ....*Systropus* Wied.

Exoprosopa.

I have tried to construct an analytical table for all the described species from the United States. In using it, it must be borne in mind that *E. gazophylax* Lw., *bifurca* L., and *agassizi* Lw.; I know only from the descriptions. All the species except *E. gazophylax* have three posterior cells.

1 (2). Style of the third antennal joint not more than one-fourth the length of that joint.................... $\begin{cases} 1. \ \textit{fasciata} \ \text{Macq.} \\ 2. \ \textit{sima} \ \text{n. sp.} \end{cases}$

2 (1). Style of the third antennal joint very long, from one-third to one-half as long as the joint, or more :
3 (14). The expanded end of the marginal cell is not altogether hyaline (more or less filled out with brown) :
4 (9). The marginal cell is altogether brown (except a small subhyaline spot in its proximal third) :
5 (6). Four submarginal cells................3. *gazophylax* Lw.
6 (5). Three submarginal cells :
7 (8). Proximal half of the wings altogether brown....4. *decora* Lw.
8 (7). Proximal half of the wings brown, but with a hyaline cross-band.................................5. *dorcadion* n. sp.
9 (4). The marginal cell contains one or more hyaline spots :
10 (13). The marginal cell contains one hyaline spot before its expanded distal end, cutting off a brown spot which fills out that end :
11 (12). Both brown cross-bands fully reach the posterior margin,
5. *dorcadion* var.
12 (11). Neither of the two cross-bands reaches the posterior margin,
7. *fascipennis* Say.
13 (10). The marginal cell contains three large hyaline spots, alternating with brown ones.........................6. *caliptera* Say.
14 (3). The expanded end of the marginal cell is altogether hyaline :
15 (20). Whole marginal cell brown, except its expanded end :
16 (17). Prevailing color of the wings brown, except the apex and a large spot in the discal cell, which are hyaline,
8. *emarginata* Macq.
17 (16). Prevailing color of the wings hyaline :
18 (19). Cross-veins at the base of the third and fourth posterior cells strongly infuscated....................:......9. *titubans* n. sp.
19 (18). Cross-veins, etc., not infuscated.............10. *dodrans* n. sp.

20 (15). Proximal half of the marginal cell hyaline :

21 (24). Proximal half of the first posterior cell brownish; its latter portion hyaline:

22 (23). Cross-veins at the base of posterior cells 1, 3, and 4 clouded with
brown .... $\left\{\begin{array}{ll} 11. & doris \text{ n. sp.} \\ 12. & agassizi \text{ Lw.} \end{array}\right.$

23 (22). Cross-veins, etc., not clouded.......... ......13. bifurca Lw.

24 (21). Proximal half of the first posterior cell hyaline, followed by a
dark brown space and then again hyaline..14. eremita n. sp.

1. EXOPROSOPA FASCIATA Macquart, Dipt. Exot., ii, 1, p. 51, 38; tab. xvii, f. 6.—E. longirostris Macquart, Dipt. Exot., suppl., 4, p. 108, probably, and Mulio americana v. d. Wulp certainly, are synonyms of this species. I also suspect that E. rubiginosa Macq. is nothing but a rubbed-off specimen of this species. E. sordida Loew, Centur., viii, 21, differs in having the anterior part of the wings darker brown, the posterior less infuscated; the base of the third and fourth posterior cells is strongly infuscated. As the habitat is Matamoras, Tamaulipas, it will probably occur in Texas.

2. EXOPROSOPA SIMA n. sp., ♀ .—Very like E. fasciata, but differs in having a shorter proboscis, which does not project beyond the oral margin, or projects very little; the whole body is more blackish; antennæ deep black; relation of the third joint to its style like 4 : 1 ; base of second and the fourth abdominal segments beset with a white, scale-like, appressed tomentum, forming cross-bands; the sides of the third segment and the whole seventh have a similar tomentum ; the pile on the sixth segment slightly yellowish; legs black; the pile on the thorax anteriorly, on the pleuræ, and above the root of the wings is pale yellowish-white or whitish-yellow, rather than fulvous ; wings like those of E. fasciata but of a more blackish-brown rather than reddish-brown color. Length 14–15ᵐᵐ.

Hab.—Humboldt Station, Central Pacific Railroad, Nevada. Three specimens, which I caught flying in the hot sunshine on the top of an arid hill (July 29).

3. EXOPROSOPA GAZOPHYLAX Loew, Centur., viii, 18.—California. I do not know this species, which will be easily distinguishable by its four submarginal cells.

4. EXOPROSOPA DECORA Loew, Centur., viii, 19.—Illinois, Iowa, Wisconsin, Colorado plains, Georgia.

5. EXOPROSOPA DORCADION n. sp.—The coloration of the wings is nearly the same as in E. caliptera. The principal difference consists in the second hyaline cross-band stopping short at the second vein, instead of reaching the first; the interval between these two veins is filled out with brown, thus connecting the two brown cross-bands, which are bifid posteriorly; the hyaline spot, which in E. caliptera exists in the marginal cell above the inner end of the second submarginal, in most cases, does not exist here; the triangular hyaline spot near the base of

the marginal cell, which in *E. caliptera* forms the anterior bifurcation of the broad brown cross-band, is much smaller here, often subobsolete.

The thorax has a fringe of reddish-brown pile anteriorly, and the usual black bristles, a stripe of white recumbent pile between the root of the wings and the scutellum, some white hairs in front of the latter; the disk of the thoracic dorsum is beset with reddish scales, mixed with white ones, the latter forming two indistinct longitudinal stripes. The abdomen has a cross-band of white scales on the anterior half of the second sedgment and a tuft of white pile at each end of this band; a small spot with white scales on the anterior margin of the third segment in the middle, and larger white spots on the posterior angles of the same segment, and two whitish scaly spots on segments 4, 5, and 6, forming two longitudinal rows, converging posteriorly; a fringe of long black pile along the sides of the abdomen, beginning with the latter part of the second segment. The ground-color of the abdomen when denuded appears as black, with red sides, the red forming indentations into the black on the hind margins of the segments. The proboscis hardly protrudes beyond the oral margin. Length 11–13ᵐᵐ.

Hab.—Seems to have a wide distribution in the Northern States, in Colorado, and in the Sierra Nevada, California. I have specimens from Summit Station, Central Pacific Railroad, California (July 17); Webber Lake, Sierra Navada, California (July 26); Shasta district, California (H. Edwards); Washington Territory (the same); Georgetown, Colorado (August 12); Twin Lake Creek, Colorado (W. L. Carpenter); White Mountains, New Hampshire (H. K. Morrison); Maine.

I said in the description that the hyaline spot in the marginal cell above the proximal end of the second submarginal *in most cases* does not exist here. There is a small spot of that kind in one of the specimens from Webber Lake; a larger one in the specimen from the White Mountains; in two specimens from Denver, Colo. (Uhler), the spot occupies the whole breadth of the marginal cell, so as to cut off the brown in its enlarged portion. I think, nevertheless, that these specimens belong to *E. dorcadion*, as their tolerably well preserved thorax and abdomen agree with the normal specimens. The corresponding hyaline spot in *E. caliptera* is not placed exactly in the same position; it is *before* the expansion of the marginal cell, while in *E. dorcadion* it is *within* that expansion.

Observation.—*E. dorcadion* is remarkably like the European *E. capucina*. As far as I can judge from the comparison of a single specimen of the latter, the wings are exactly the same, but there seems to be a difference in the distribution of the white scales on the abdomen. This resemblance has given rise to a confusion which may provoke a discussion about the true specific name of *E. dorcadion*. There is no doubt that this species is the true *Anthrax capucina* of Fabricius, marked "habitat in America boreali" (see Syst. Antl., p. 123, 23). Wiedemann, observing the resemblance of Fabricius's types to the European species, concluded too hastily that the *habitat* assigned to them by Fabricius was erroneous,* and transferred the name to the European species; all the European authors have followed his example since. The zealots of priority will probably insist upon changing the name of the European species, now adopted in all the works on European *Diptera ;* in my opinion, it is much more in accordance with the true interest of science to preserve a name which has been so long in use, and merely to strike out the quotation from Fabricius. Furthermore, it may turn out in the end that *E. dorcadion* is the same species as the European *E. capucina*, in which case there will be no occasion for a discussion.

6. EXOPROSOPA CALIPTERA Say, Journ. Acad. Phil., iii, 46 (Compl. Writings, ii, 62).—To Say's very clear description, I will add a statement about the silvery spots on the abdomen, taken from two well preserved specimens in my possession. A silvery cross-band on the second segment occupies two thirds of the breadth of the segment, and is deeply emarginate in the middle; a silvery spot on the posterior corners of the third segment, and *a silvery longitudinal streak in the middle of each of the segments 4, 5, and 6.* The latter character is important, as it does not exist in *E. dorcadion,* otherwise so closely allied. I caught this species near Cheyenne, Wyo., August 21, and also received a specimen from Morino Valley, New Mexico, collected by Lieut. W. L. Carpenter, July 1.

Observation.—There is an *Anthrax caloptera* Pallas (see Wiedem., Zool. Magaz., i, 2, 12, and Meigen, Syst. Beschr., ii, 173), which Wiedemann considered the same as *A. capucina* Fab., and therefore put among the synonyms. As the name will probably never be revived, Say's name may be retained.

Walker's *Anthrax californiæ* agrees better with this species than with *E. dorcadion.*

7. EXOPROSOPA FASCIPENNIS Say is well known. *Anthrax noctula* Wiedemann (Auss. Zw., ii, 635, 45) and *Exopr. coniceps* Macquart (4e suppl., 108) are its synonyms. *Exopr. phila delphica* is, I suspect, only a smaller variety, which occasionally occurs.

8. EXOPROSOPA EMARGINATA Macquart, Dipt. Exot., ii, 1, 51.—Virginia, Georgia, not rare.

9. EXOPROSOPA TITUBANS n. sp.—Head grayish black; antennæ black; style of the third antennal joint nearly as long as the joint; face beset with a golden-yellow scaly tomentum; oral margin reddish; front with some golden scaly hairs and black pile; posterior orbit with short, appressed, white hairs; proboscis not protruding. Thorax black, with dingy-yellowish pile; scutellum reddish, with yellow pile. Abdomen black in the middle, reddish on the sides; anterior half of the second and the fourth and sixth segments with a cross-band of white scales; the latter half of the second and the other segments beset with yellow-

---

* The passage from Wiedemann's Zool. Mag., i, 2, 12, is quoted by Meigen, Syst. Beschr.,

ish scales (somewhat rubbed off in my specimens); some black pile in the posterior corners of the segments, beginning with the second. Venter reddish, sparsely beset with yellowish scales. Legs black ; femora beset with fulvous scales. Wings hyaline, brown anteriorly, which color is bounded by the fourth vein before the small cross-vein and by an oblique line running to the end of the first vein after it; cross-veins at the base of the third and fourth posterior cells with broad brown clouds. Length 12–13^{mm}.

Hab.—Denver, Colo. (P. R. Uhler); one female. A second specimen, also from Denver (by the same), is a little larger; the proboscis is protruding a little beyond the oral margin ; the brown of the wings is darker and encroaches considerably on the second basal, and also a little on the anal cells; the posterior femora are more densely clothed with fulvous scales ; the two last posterior cells are longer. I am in doubt whether to consider this a distinct species or not. The wings of this species must be very like those of *E. sordida* Loew ; but the latter must have a longer proboscis, and the antennæ, I suppose, must be like those of *E. fasciata,* to whom Dr. Loew compares it,—that is, the third joint must be three or four times longer than its style, and not nearly equal to it in length.

10. EXOPROSOPA DODRANS n. sp.—The brown of the anterior part of the wing is bounded by the basal cross-veins, by the fourth vein as far as the small cross-vein, beyond which the boundary-line runs obliquely toward the end of the first vein ; the second basal cell at its proximal end is considerably encroached upon by the brown ; the first posterior and first submarginal cells likewise ; the remainder of the wing is hyaline, without any spots or clouds, except an indistinct one on the cross-vein at the base of the fourth posterior cell. Head black, clothed with golden-yellow short pile ; oral margin reddish ; antennæ black; first joint red ; third joint conical, with a style half as long as the joint ; proboscis projecting about half the length of the head beyond the oral margin. Thorax grayish-black, with yellowish pile, more whitish above the root of the wings ; scutellum reddish, black at base ; abdomen black, with very little red on the sides ; second segment with the usual white cross-band on its anterior half ; the other segments beset with yellow and white scales; sides with yellowish hair at the base and black hair on the segments beyond the second ; venter reddish, with white scales on the first four, and with yellow ones on the following segments. Halteres brownish at the base of the knob ; its tip yellowish. Legs densely beset with fulvous scales, which cover the ground-color ; tarsi black. Length 12^{mm}.

E. dodrans is, in many respects, very like *E. titubans;* it is, however, a little smaller ; the brown on the wings is of a very uniform tinge, with no reddish nor subhyaline spots in it. In *E. titubans,* the brown is more reddish, the costal cell more yellowish, and the proximal end of the submarginal cell has a paler, almost subhyaline, spot in it ; the clouds on the cross-veins at the base of the third and fourth posterior cells are very distinct here, while in *E. dodrans* there is a hardly perceptible infuscation on the cross-vein at the base of the fourth cell only. In *E. dodrans*, the distance between the bases of the third and fourth posterior cells is a little greater than in *E. titubans*; finally, in the latter, the antennal style is comparatively longer.

Hab.—Colorado Springs (P. R. Uhler); two specimens (somewhat rubbed off on the abdomen). One of them has, on both wings, an adventitious stump of a vein inside of the discal cell.

11. EXOPROSOPA DORIS n. sp.—Base of the wings as far as the basal cross-veins brownish; costal cell yellowish; first basal cell, except its proximal end, which is pale yellow, and first posterior cell, except its distal end, brown; the middle portion of the marginal and first submarginal cells brown, which thus forms an incomplete, irregular, and ill-defined cross-band, expanded anteriorly as far as the end of the first vein, attenuated posteriorly, and ending in the brown of the first posterior cell; the distal boundary of this cross-band is in zigzag, one of the projections touching the proximal end of the second submarginal cell; the proximal boundary is evanescent; a round, brown spot on the proximal end of the second posterior cell; an irregular, ill-defined, narrow, oblique, brown band runs from the small cross-vein across the discal cell, covers the proximal end of the third posterior cell and the posterior cross-vein, cuts in two the anal cell, and ends in the axillary without touching the posterior margin; small clouds on the bifurcation of the second and third veins and on the proximal end of the discal cell. Epistoma yellow, clothed with yellow scales; cheeks pale yellow, with a silvery covering of scales; front and vertex black, with golden-yellow scales; posterior orbits silvery; proboscis not projecting beyond the oral margin; antennæ black; first joint short, reddish; the third conical, with a style half as long as the joint. Thorax with yellow pile; white pile above the root of the wings and in front of the reddish scutellum; silvery-white pile on the chest and pleuræ; abdomen densely clothed with yellow scales, except at the base of the second and on the fourth segments, where there are cross-bands of white scales; seventh segment likewise beset with white scales; venter reddish-yellow, with snow-white scales on the first four segments and yellowish scales on the following segments, with an admixture of black ones on the fifth segment. Femora red, clothed with fulvous scales; tibiæ reddish, darker on their front side; the front pair black at tip; the hind pair is black, thickly clothed on the inner side with fulvous scales; tarsi black. Halteres with a yellow knob. Length 7–8^{mm}.

Hab.—Humboldt Station, Central Pacific Railroad, Nevada (July 29). A single very well preserved specimen.

A second specimen, from Oregon (H. Edwards), is considerably larger from 12^{mm} to 13^{mm}); the coloring of the body is exactly the same; the distribution of the brown spots on the wings is, in the main, the same, but

4 H B

they are all less extended and weakened in intensity of color; the whole first basal and the proximal end of the first posterior cell are not brown, but yellowish, which color is interrupted by a brown cloud on the small cross-vein, and ends in a brown cloud in the middle of the first posterior cell; the brown band across the middle of the marginal and first submarginal cells is narrower; that running obliquely from the discal to the axillary cell is likewise narrow, almost dissolved in its component spots. I am inclined to believe, nevertheless, that it is the same species.

12. EXOPROSOPA AGASSIZI Loew, Centur., viii, 24.—California. Must be somewhat like *E. doris;* nevertheless, a different species.

13. EXOPROSOPA BIFURCA Loew, Centur., viii, 23.—California. I do not know it.

14. EXOPROSOPA EREMITA n. sp.—Wings brown at base, the brown encroaching a little beyond the basal cross-veins, and with two broad brown cross-bands; the first is limited anteriorly by the præfurca and ends in the distal half of the axillary cell, where a very narrow hyaline space separates its end from the margin of the wing; the second starts from the anterior margin in the region of the stigma, and, attenuated posteriorly, stops short before crossing the third posterior cell; the yellowish-brown costal cell forms the only connection between those three regions of brown, the hyaline intervals between which are almost broader than the brown cross-bands; apex of the wing and posterior margin likewise hyaline. Front and vertex black, beset with yellowish pile; epistoma brownish-red; antennae black, third joint conical, with a style about one-third as long as the joint. Proboscis hardly projecting. Thorax grayish-black, beset with yellowish pile; antescutellar callosities brownish; scutellum reddish-brown, black at base. Ground-color of the abdomen black, with red sides; second segment with a white cross-band at the base; white spots on each side of the third, and interrupted cross-bands on the fourth and fifth; sixth segment also whitish; yellow pile on the sides of the abdomen, at the base, and black pile beyond the end of the second segment. Venter red, with traces of a covering of white scales on segments 2–4. Legs dark reddish-brown, with black pile. Length 16^{mm}.

Hab.—Shasta district, California (H. Edwards). A single specimen. Its epistoma and abdomen were somewhat denuded.

DIPALTA nov. gen.

Differs from *Exoprosopa* in the course of the second vein, which is strongly contorted, in the shape of a recumbent S, near its point of contact with the cross-vein, which separates the first submarginal cell from the second.

A still more important difference lies in the structure of the antennæ, the third joint of which does not bear the terminal style, so apparent in *Exoprosopa*, and is more like that of the genus *Anthrax*. Examined attentively, that joint shows, on its incrassate, basal part, a more or less distinct transverse suture, which also exists in *Anthrax*, and may indicate that the third joint is very much shortened here and coalescent with the style, their suture being very near the base of the joint. The joint is onion-shaped at the base, with a slender, gently tapering, almost linear, prolongation, ending in a point. I do not see any bristle at the end. The body is more slender than in *Exoprosopa*, and the hairy covering of a more uniform color.

Anthrax paradoxa Jaennicke (Exot. Dipt., 31, tab. ii, 16) probably belongs here. The course of the second vein is the same in that species; only the cross-vein in *D. serpentina* is inserted in the middle of the sinus, and not at its base. In the specimen from Georgia, to be mentioned below, the cross-vein is placed very nearly as it is in Jaennicke's figure.

Diplocampta Schiner (Novara, 119, tab. ii, f. 9) from Chili resembles *Dipalta* in the curvature of the second vein; nevertheless, it is evidently different; the abdomen is not longer than the thorax, and narrower, being gradually attenuated posteriorly; the structure of the antennæ is different; the species is small (1½ to 2 lines long), with nearly hyaline wings; the position of the cross-vein between the second and third veins is different.

Dipalta, in Greek, means *twice bent*.

1. DIPALTA SERPENTINA n. sp., ♂ ♀.—Body black, densely and rather uniformly clothed with a short, appressed, pale yellow tomentum; longer pile on the pleuræ and on the anterior margin of the thorax. Wings subhyaline, with a pale brownish-yellow tinge, the base and two irregular cross-bands brown. These cross-bands are formed by the confluence of large, round, brown spots on the cross-veins and bifurcations; the first cross-band has a hyaline spot anteriorly in the proximal half of the marginal cell; posteriorly it is attenuated; the second cross-band contains a hyaline spot at its posterior end in the first and second posterior cells; the apex of the wing is hyaline, the end of the second vein clouded with brown; a much smaller cloud on the third vein; a stump of a vein projects into the marginal cell a short distance before the cross-vein; beyond the cross-vein, the second vein is very deeply bisinuate. Legs densely clothed with fulvous scales, except on the tarsi, which are black. Antennæ, first two joints very short, reddish, sometimes darker; third joint black. Halteres yellow, slightly brownish at the base of the knob. Proboscis not projecting beyond the oral border. Length 10-11ᵐᵐ.

Hab.—Mount Shasta district, California (H. Edwards); Clear Creek Cañon (Uhler); Colorado Mountains (W. L. Carpenter). Three specimens.

Observation.—I have a specimen from Georgia (H. K. Morrison) which probably belongs here. The hairy covering of the body is of a very saturate fulvous; the markings of the wings are of a darker brown; the different position of the cross-vein has been alluded to above.

ANTHRAX.

In this genus, I abstain almost entirely from describing the numerous new species which I have from California and the Colorado region. Such a work should be done in connection with a critical examination of the previously described species from the Atlantic States, and my material from those States is entirely insufficient for that purpose. I subjoin a list of all the described species from North America north of Mexico. The rough grouping which I have attempted will facilitate, I hope, the task of identification.

A. *Wings hyaline; costal cell but little darker:*
    albipectus Macq., 3e suppl., 34.—North America.
    *alternata Say, Compl. Wr., ii, 61.—Middle States.
      ? SYN.—*consanguinea* Macq., D. E., ii, 1, 69.
    *connexa Macq., 5e suppl., 76.—Baltimore.
    gracilis Macq., D. E., ii, 1, 76.—Philadelphia.
    hypomelas Macq., D. E., ii, 1, 76.—North America.
    *lateralis Say, Compl. Wr., ii, 59.—Middle States.
    mucorea Lw., Centur., viii, 43.—Nebraska.
    *scrobiculata Lw., Centur., viii, 39.—Illinois.
    stenozona Lw., Centur., viii, 40.—Illinois.
    molitor Lw., Centur., viii, 42.—California.

B. *Wings hyaline; costal cell dark brown:*
    *fulviana Say, Compl. Wr., i, p. 253.—Northwestern Territory (Say).
    * nigricauda Lw., Centur., viii, 38.—Massachusetts; White Mountains.

C. *Wings with the proximal half dark brown or black (like* FULVOHIRTA, SINUOSA):
    albovittata Macq., 4e suppl., 113.—N. A. (?).
    * celer Wied., i, 310.—Kentucky.
    edititia Say, Compl. Wr., ii, 353.—No locality given.
    floridana Macq., D. E., 4e suppl., 112.—Florida.
    * fulvohirta Wied., i, 308.—Atlantic States.
      SYN.—*conifacies* Macq., 4e suppl.
        separata Wk., Dipt. Saund., 177.
    (?) *incisa* Walker, Dipt. Saund., 187.—North America.
    (?) *cedens* Walker, Dipt. Saund., 190.—United States.
    palliata Lw., Centur., viii, 32.—Illinois.
    parvicornis Lw., Centur., viii, 36.—Illinois.
    sagata Lw., Centur., viii, 34.—Matamoras.
    * sinuosa Wied., i, 301.—United States.
      SYN.—*concisa* Macq., D. E., ii, 1, 68.
        nycthemera Macq., D. E., ii, 1, 67.
        (?) *assimilis* Macq., suppl., i, 114.
    vestita Walker, List, ii, 258.—Nova Scotia.
    curta Lw., Centur., viii, 35.—California.
    diagonalis Lw., Centur., viii, 33.—California.

D. *Species resembling* ANTHRAX HALCYON *Say:*
   * *ceyx* Lw., Centur., viii, 30.—Virginia.
     demogorgon Wk., List, ii, 265.—Florida.
   * *flaviceps* Lw., Centur., viii, 29.—Tamaulipas.
   * *halcyon* Say, Compl. Wr., i, 252.—Northwestern United States.
     fuliginosa Lw., Centur., viii, 31.—California.
   * *alpha* n. sp.—California; Cheyenne, Wyo.
E. *Species resembling* ANTHRAX TEGMINIPENNIS *Say:*
   * *lucifer* Fab., Wied., i, 294.—West Indies; Texas.
     SYN.—*fumiflamma* Walk., D. S., 184.
   * *tegminipennis* Say, Compl. Wr., i, 253.—Northwestern Territory,
     (Say).
     fuscipennis Macq., H. Nat., i, 410. (Very doubtful species.)
F. *Peculiar species not coming within the former groups :*
     pertusa Lw., Centur., viii, 28.—Pecos River, Western Texas.
     (Wings bifasciate with brown.)
     proboscidea Lw., Centur., viii, 27.—Sonora. (Very long proboscis.)

ANTHRAX SINUOSA Wiedemann, Auss. Zw., i, 301, 64.—I have specimens of this widespread species from Sonoma County, California (July 5), Vancouver Island (H. Edwards), Clear Creek Cañon, Colorado (P. R. Uhler), Manitou, Colo. (August 17), Morino Valley, New Mexico (W. L. Carpenter). The Californian specimens have the short, scale-like hairs covering the thorax and the abdomen of a more intense color, red instead of fulvous. In all the western specimens, the brown in the first posterior cell reaches the bifurcation of the third vein, sometimes even beyond. The brown spot at the distal end of the first submarginal cell is a little larger, so as to encroach a mere trifle on the second submarginal. This species has small but very distinct pulvilli, and thus holds the middle between *Anthrax* and *Hemipenthes*.

ANTHRAX HALCYON Say.—Easily distinguished by its third posterior cell being in most specimens bisected by a cross-vein. It has a wide distribution in the Northwest and West; it is not rare round Manitou and Colorado Springs, Colorado, in July and August. I also have it from Morino Valley, New Mexico, July 1 (W. L. Carpenter).

ANTHRAX FULIGINOSA Loew.—Among my specimens from the West, longing to the group of *A. halcyon*, there are several species; three at least occur in California. I am not sure whether *A. fuliginosa* is among them.

The following species, which seems to occur on the plains of Wyoming as well as in the Sierra Nevada, is easily distinguishable from all the described species:

ANTHRAX ALPHA n. sp., ♂ ♀.—Coloring of the wings very like that of *A. halcyon;* second submarginal cell bisected by a cross-vein; in the third posterior, a long stump of a vein. Length 12–14ᵐᵐ.

Front, face, and cheeks beset, the former with short black, the two latter with scarce yellow pile; vertex black; proboscis not projecting; antennæ black, first joint reddish, with black pile. Thorax grayish-black, clothed with pile, which is pale fulvous above and white on the pectus and the lower part of the pleuræ. Scutellum reddish, black at base; ground-color of the abdomen is grayish-black ; in rubbed-off specimens, only a little red is perceptible on the sides of the second and third segments; in well preserved specimens, the ground-color is entirely concealed under a dense, appressed tomentum, which is whitish-gray on the anterior and brownish-fulvous on the posterior half of the segments; an ill defined blackish spot in the middle of each segment; the sides of the first two segments are beset with yellowish-white pile; the sides of the following segments, beginning with the end of the second, with black, mixed with fulvous pile, the black forming tufts on the hind margins of the segments ; the same black pile is scattered over the surface of the abdomen, above the tomentum. Venter: segments 2–4 reddish, more or less black at the base; the following segments black, with a reddish posterior margin. Legs red, with a golden-yellow tomentum and black spines ; front femora black at base ; tips of tibiæ and all the tarsi black. Wings tinged with blackish-brown ; in the apical half, the following spaces are grayish-hyaline : a spot in the expanded end of the marginal cell, the end of the first submarginal and nearly the whole second submarginal, a streak in the end of the first posterior cell, the three other posterior cells, and the latter half of the discal cell ; the veins traversing these subhyaline places are clouded with brown. The cross-vein bisecting the second submarginal cell is placed in its narrow part, so as to form with the adjacent veins the figure A.

Hab.—Cheyenne, Wyo., where I found it to be quite common, August 21, 1876. Five specimens.

Six specimens which I took near Webber Lake, Sierra Nevada (July 25), agree in all respects with those from Cheyenne ; but they are a little smaller, the coloring is a little darker, both on the wings and on the body; the pile on the chest and pleuræ is less white; the tomentum on the abdomen above is the same, but the fulvous prevails over the gray, and the black spots in the middle of each segment are larger; on the second segment, along the hind margin, the black forms a cross-band, attenuated on each side, and not reaching the lateral margin ; the same is repeated on each following segment, the black spot rapidly diminishing in extent. The venter is reddish, without any black at the base of the segments. The portion of the anterior branch of the third vein beyond the supernumerary cross-vein is very distinctly clouded with brown in these specimens, while it is not clouded at all, or only imperceptibly, in the specimens from Cheyenne. I hold this to be merely a local variety of *A. alpha.*

ANTHRAX LUCIFER F.—Hitherto known only from the West Indies. I have several specimens from Dallas, Tex. (Boll), which seem to belong here.

I have one or two other well marked species, belonging to the same group; one, from Fort Bridger, Wyo., is somewhat like *A. lucifer*, but certainly different; the other, from South Park, Colorado, and Twin Lake Creek, Colorado (W. L. Carpenter), is more like an *A. tegminipennis*, but with a more hyaline latter half of the wings.

HEMIPENTHES.

This genus was established by Loew (Centur., viii, 44) for the European *Anthrax morio*, and for a species from British North America, of which he described the female as *H. seminigra*. He does not say what prevented him from identifying it with *Anthrax morioides* Say, which is certainly a *Hemipenthes*. A well preserved male specimen from Montreal agrees well with Say's and Wiedemann's descriptions; the knob of the halteres, however, is dark on one side, whitish-yellow on the other. It does not have the strongly coarctate first posterior cell, which distinguishes *H. seminigra*, according to Mr. Loew's description. A number of specimens from Spanish Peaks, Colorado, possess this character, and therefore belong, I suppose, to *H. seminigra*.

I have, moreover, a number of specimens from different parts of California (Yosemite and Webber Lake), which may likewise belong here. The *Hemipenthes* occurring commonly in Marin County, California, in May, seems to be a different species. The wings are more uniformly black, and the pile and tomentum on the abdomen are different.

ARGYRAMŒBA.

Contains a variety of forms, which have, as common characters, a pencil of hairs at the end of the third antennal joint, and distinct, rather large pulvilli. With a very insufficient material, I have attempted a rough grouping of all the described species from the United States, in which I have used some characters hitherto neglected. In using this table, it must be remembered that I know only those species which are marked with an asterisk (*), and that data about the others are drawn from the descriptions. I have omitted *Anthrax costata* Say (Compl. Wr., i, 254), which may possibly be an *Argyramœba*.

I. Large species; third posterior cell bisected by a cross-vein :
   * *simson* Fab., Wied., i, 259 (syn. *scripta* Say, Compl. Wr., ii, 59).—Atlantic States.
   delila Loew, Centur., viii, 45.—California.
II. The male has the last abdominal segments clothed with silvery scales; the other segments in both sexes altogether black :
   (a) Wings black; posterior margin hyaline; the limit of the black well defined :

(*b*) The black of the anterior margin of the wing reaches the end of the first submarginal cell; wings remarkably narrowed toward their root, cuneiform; axillary cell exceedingly narrow, linear; alula obsolete; the pencil-bearing style of the antenna is as long as the styliform portion of the third joint:

argyropyga Wied., i, 313 (syn. *contigua* Lw., Cent., viii, 50, *female*).—Virginia; Georgia.

(*bb*) The black on the anterior margin of the wing does not cover the end of the marginal cell; wings broad, with a fully-developed, broad axillary cell and alula; the pencil-bearing style of the antenna is short, about one-quarter of the length of the very long styliform portion of the third joint:

analis Say, Compl. Wr., ii, 60 (syn. *georgica* Macq.).— Atlantic States.

(*aa*) Wings black, this color being gradually evanescent posteriorly; they are long and comparatively narrow; the stout basal portion of the third antennal joint is somewhat gradually attenuated, conical, with a short pencil-bearing style.

cephus Fabr., Wied., i, 297.—Southern States.

III. Hind margins of abdominal segments (especially the second and third) more or less beset with white or whitish scales, forming spots or cross-bands:

(*a*) Basal portion of the wings more or less black; black spots at the proximal end of the second submarginal, the third posterior, sometimes also the second posterior cells:

(*b*) A hyaline space across the proximal end of the marginal and the distal end of the first basal cells:

(*c*) A brown spot at the proximal end of the second posterior cell:

limatulus Say, Compl. Wr., ii, 354.—Indiana (Say).

(*cc*) No brown spot, etc.:

albofasciata Macq., D. E., ii, 1, 67.—Georgia.

(*bb*) No hyaline space across, etc.:

(*d*) Distal half of the wings hyaline:

antecedens Walker, Dipt. Saund., 193.—United States.

(*dd*) Distal half of the wings distinctly grayish:

obsoleta Loew, Centur., viii, 47.—Missouri; Georgia.

(*aa*) Basal portion of the wings black, with numerous black spots in the hyaline portion:

pluto Wied., i, 261.—Kentucky (Wied.).

stellans Loew, Centur., viii, 46.—Oregon.

(*aaa*) Basal portion of the wing more or less dark; no black spots at the proximal end of the second submarginal, second and third posterior cells:

bastardi Macq., D. E., ii, 1, p. 60.—North America.

pauper Loew, Centur., viii, 48.—Illinois.

**fur* n. sp.—Texas.

IV. Group in which the costal cell is checkered, hyaline, and black:

**œdipus* Fab., Wied., i, p. 262 (syn. *irrorata* Say and Macq.).

Schiner (Fauna Austr., Dipt., i, 52) says that the larvæ of *Argyramoeba* live parasitically in pupæ of *Lepidoptera*. That this is far from being universally the case is proved by the fact that *A. cephus* and *A. fur* were bred from the nest of a Mud-wasp in Texas, forming tubes of clay five or six inches long, pasted together like organ-pipes. The nests were found near Dallas, Tex., by Mr. Boll, and are now in the Museum of Comparative Zoölogy in Cambridge, Mass. The pupæ bored their way directly through the clay, and the exuviæ remained in the hole. The Hymenopteron which builds these nests is very probably a *Pelopæus;* the larva of the fly probably devours the larvæ of the wasp.

I observed *A. œdipus* in the Sierra Nevada persistently flying round a hole in a pine log, probably containing the nest of some Hymenopteron.

Argyramœba leucogaster Meig. was bred from the nest of a *Cemonus*, living in deformed reeds. The article of Mr. Frauenfeld on the subject is well worth reading (Verh. zool.-bot. Gesellsch., 1864, 688).

A. subnotata Meig. was bred by the same author from a nest of *Chalicodoma muraria* Lin. (Verh. zool.-bot. Ges., 1861, 173).

A. sinuata was bred by Mr. Laboulbène from the nest of a Hymenopteron, probably *Megachile muraria* (see Ann. de la Soc. Entom. de France, 1857, 781).

1. ARGYRAMŒBA ŒDIPUS Wied. (syn. *irrorata* Say, Macq.).—Seems to have a very wide distribution all over North America, even quite far in the northwest of the British possessions; according to Schiner, also in South America. I brought a couple of specimens from Webber Lake, Sierra County, California. A specimen from the Shasta district was given to me by Mr. H. Edwards. A specimen which I took in Sonoma County, July 4, is larger, and the black dots in the latter part of the wing are much more scarce.

2. ARGYRAMŒBA LIMATULUS Say.—I retain under this name a group of specimens from the Geysers, Sonoma County, California (May 5–7); Fort Bridger, Wyoming (August 7); Fair Play, Colorado; Spanish Peaks, Colorado; Sangre de Cristo Mountains, New Mexico (the latter collected by Lieut. W. L. Carpenter). The extent of the black on the wings in these specimens is very variable, even in those taken on the same day and in one locality; in many, there is very little black left except the dark clouds on the cross-veins. Whether these specimens really belong to the *A. limatulus* I am not prepared to affirm. Say's orig

inal specimens were from Indiana. I have no specimens from that region, except a well preserved male from Detroit, Michigan, which, besides the white scales, has some fulvous scales forming cross-bands on the abdominal segments. Such scales not being mentioned by Say, I am in doubt whether this specimen is his *limatulus* or not.

3. ARGYRAMŒBA PLUTO Wied.—The basal half of the wing is more or less like the darker-colored specimens of *A. limatulus;* the hyaline portion has nine or ten small black spots, one at the extreme end of the first vein in the shape of a small cloud, two on the concave end of the second vein, two on the anterior branch of the third vein, the one at its origin being large; a spot, sometimes double, on the cross-vein at the base of the second posterior cell, often coalescent with a small spot on the vein separating this cell from the third posterior; a spot on the curvature of the cross-vein at the base of the third posterior cell; another at the proximal end of that cell; one a little before the tip of the fifth vein. The large spot on the cross-vein at the base of the fourth posterior cell is usually coalescent with the black on the anterior half of the wing. A long stump of a vein on the geniculate base of the second vein; two stumps on the sinuosities of the anterior branch of the third vein, one on each side; a small stump on the cross-vein at the base of the third vein. Body deep black; face and front with short, erect, black pile, mixed with some white ones, especially around the mouth. Abdomen with tufts of black pile on each of the first segment and some white hairs along its hind margin; small patches of white scales on the sides and the hind margin of the third and fourth segments; the end of the abdomen in the male densely beset with white scales.

This description applies to specimens about 11^{mm} long, which I have from Canada (F. X. Bélanger), Pennsylvania (E. T. Cresson), and Waco, Texas (Belfrage). But I have two other specimens, from Illinois (Le Baron) and Texas (M. C. Z.), in which the black spots on the distal half of the wings are so much enlarged that they coalesce and form an irregular, broad cross-band, bifurcate at both ends; only three brown dots on the apex of the wing are not confluent with this cross-band. A specimen from Georgia (H. K. Morrison) holds the middle between these two forms, its spots being larger than in the first form, and less coalescent than in the second. This last specimen, as well as that from Illinois, measures only 8–9^{mm}. I have little doubt now that all these specimens belong to *A. pluto.*

5. ARGYRAMŒBA FUR n. sp., ♂ ♀.—Face and front pollinose with yellowish-gray, and clothed with short black pile; on the front, some very minute pale yellow hairs are mixed with the gray ones; the gray occiput is more or less clothed on the orbits with hair of this latter description. The dull grayish-black ground-color of the thorax and scutellum is in part covered with a short, yellow, appressed tomentum; a large tuft of yellowish-white pile between the humerus and the root of the wings, extending to the upper part of the pleuræ; a frill of whitish

hairs, mixed with black ones, on the anterior margin of the thorax, opposite the occiput; the edge of the scutellum with a yellowish-white tomentum and a row of stiff black bristles. Ground-color of the abdomen black; first segment on each side with a tuft of yellowish-white pile and a sparse fringe of them along the hind margin; second segment black, with a faint streak of microscopic fulvous tomentum in the middle; the following segments are densely clothed with a recumbent, short, yellowish tomentum, more whitish on the hind margins of the segments; in the middle of each segment, the tomentum, being less dense, leaves a dark spot, which, in connection with similar spots on the next segments, forms an ill-defined longitudinal dark stripe; rows of black erect pile on each segment above the yellow pubescence; the black pile is more dense on the sides and at the end of the body; the sides of the two last segments are clothed with whitish, scale-like pile; the same whitish scales form subtriangular spots on the hind corners of the third and fourth segments, connected with the fringes of whitish hairs on the hind margins of the segments. Femora black; four front tibiæ and tarsi dark brown; the front femora are sparsely beset on the anterior side with whitish-gray scales. Halteres yellow, the knob with a brown spot. Wings grayish-hyaline; their root, the costal cells, the two basal cells, and the proximal ends of the anal and axillary cells pale reddish-brown; a transverse darker brown cloud on the small cross-vein, and the bifurcation of the second and third veins; another darker cloud, coalescent with the brown of the base of the wings, lies between the origin of the præfurca and the cross-vein at the base of the fourth posterior cell. The stump of a vein on the anterior branch of the third vein is small, sometimes obsolete; that on the curvature of the second vein is moderately long; there is none in the second posterior cell. Length 10–11^{mm}.

Hab.—Dallas, Texas (Boll). Three specimens. The description was drawn from a well-preserved female.

As mentioned in the introductory paragraph to this genus, the larva of this species lives in the nest of a Mud-wasp (*Pelopœus?*), and this alone induced me to describe it. I did not dare to identify it with *Anthrax bastardi* Macq., and still less with *Anthrax costata* Say.

TRIODITES nov. gen.

Belongs in the number of genera which forms the passage between the *Anthracina* and *Bombylina*. It has the appearance of an elongated *Anthrax*; the antennæ are distant at the base, and resemble those of that genus. But the eyes of the male are contiguous for a short distance, high up, on the vertex, and the bifurcation of the second and third veins takes place long before the small cross-vein, the præfurca being very short. *Triodites*, like *Oncodocera* and other genera of the same group, has, in the male, a very large frontal triangle, and, in the female a correspondingly large frontal space.

Head subglobular; *occiput* but slightly tumid; *oral opening* oval, slightly oblique; *front* projecting very little in the profile; epistoma re-treating below the antennæ, comparatively long (longer than in *Anthrax*) and broad, nearly flat; cheeks exceedingly narrow, linear, the eyes being nearly in contact with the mouth; *eyes* very large, occupying the whole side of the head, and descending on the under side to the very edge of the mouth; they are (as in *Anthrax*) somewhat reniform, with a linear impression, starting from the sinus of the occipital orbit and interrupted about the middle; in the male, the eyes come in contact for a short distance in front of the small ocellar triangle; in the female, they are separated by an interval, which is not greater than the interval between the roots of the antennæ; frontal triangle, in the male, very large, nearly flat; the pubescence on front and epistoma is short, denser on the latter than on the former.

Antennæ, in a profile-view of the head, are inserted about its middle; the two basal joints are exceedingly short, concealed in the pubescence; the third joint, broad at base, becomes suddenly contracted, long, linear, styliform, truncate at the end, the truncature bearing a minute joint, with a bristle at the end.

Thorax rounded, clothed with a very delicate, even, silky, erect, and moderately long pubescence; only a single delicate bristle is perceptible on each side in front of the wings, and a few on the antescutellar tubercle (they are of the same color with the pubescence); *scutellum* rather broad.

Abdomen, in the male, narrower than the thorax, and not much longer, cylindrical, the seventh segment being only a little narrower; in the female, the abdomen is a little longer than the thorax, and nearly as broad, gradually attenuated posteriorly. The segments do not differ much in length, the second being but a little longer; in the female, the seventh, at the end, bears a dense circle of appressed hairs, their ends converging, and closing the anal opening.

Legs moderately long, clothed with scales, and beset with spines; pulvilli distinct.

Wings of moderate length and breadth, narrower than in most species of *Anthrax;* venation like that of an *Anthrax*, except that the bifurcation of the second and third veins takes place very early, at the same distance from the root of the wing as the proximal end of the discal cell; the præfurca is but one-half longer than the great cross-vein; the small cross-vein is about the middle of the discal cell; the curvature of the second vein at the end and of the anterior branch of the third vein are very much like those of an ordinary *Anthrax* (*A. alternata* or *sinuosa*); costal enlargement small; a distinct, apparently coriaceous, epimeral hook, as in *Anthrax*.

Triodites, in Greek, means a *street-lounger*.

TRIODITES MUS n. sp., ♂ ♀.—Uniformly clothed with whitish-gray pile; face with white pile; wings hyaline. Length 8–9^{mm}.

Male.—Frontal triangle black, with short, erect, black pile; face with a dense covering of short snow-white pile; antennæ black; occiput black, with appressed white hairs along the orbits; thorax grayish-black, with a dense covering of delicate, downy, whitish-gray pile, which in an oblique light looks altogether white; the few bristles on the ante-scutellar callosities and on the scutellum are whitish, almost colorless; abdomen black, with the same covering of grayish-white pile, which is longer here on the sides. Halteres whitish; knob brownish. Legs black, densely clothed with white scales; spines on femora and tibiæ whitish-yellow. Wings, including the costal cell, of a pure hyaline; veins, except at the root, black; costal and first longitudinal brown.

Female.—Like the male, but the front is slightly brownish-pruinose, and has, besides the erect, black pile, some short, recumbent, yellowish hairs. The hind margins of the abdominal segments are beset with some short, appressed, whitish hairs, forming cross-bands.

Hab.—I have a single male, which I took near the Salt Lake, Utah, August 1. One of the females is from Sonoma County, California, July 6; the other from the Shasta district (H. Edwards, July, 1875).

ONCODOCERA.

ONCODOCERA LEUCOPROCTA Wiedemann, i, 330 (*Mulio*); male (syn. *Oncodocera dimidiata* Macq., D. E., ii, 1, 84, female).—Middle and Southern States; Illinois; Wisconsin.

LEPTOCHILUS MODESTUS Loew, Centur., x, 40.—Texas.) I do not know
APHŒBANTUS CERVINUS Loew, Centur., x, 39.—Texas.) these species

BOMBYLIUS.

The species from the Atlantic States may be grouped as follows :—

1. Anterior half of the wings brown, with a well-defined, sinuous, posterior limit of that color :
    fratellus Wied., i, 583.
    > SYN.—*vicinus* Macq., ii, 1, 98.
    >   *albipectus* Macq., 5e suppl., 82.
    >   *æqualis* Harris (*nec* Fabr.), 3d edit., 604.
    >   *major* Kirby, Fauna Bor.-Am., 312.

2. Wings brown at base, and with brown spots on the posterior half :
    pulchellus Loew, Centur., iv, 47.
    pygmæus Fab., Wied., i, 351.

3. The brown of the basal half of the wings is gradually evanescent posteriorly :
    atriceps Loew, Centur., iv, 49.
    mexicanus Wied., i, 338.
    > SYN.—*fulvibasis* Macq., 5e suppl.   ) Synonymy on the authority of Mr.
    >   *philadelphicus* Macq., ii, 1, 99.)   Loew, in litt.
    rarius Fab., Wied., i, 335.
    validus Loew, Centur., iv, 48.

The Californian species which I have before me may be tabulated as follows:

1. Anterior half of the wings brown, which color has a well-defined posterior border ,.......................1. *major* Linn. ∨
2. Anterior part of the wings brown; the posterior, or hyaline region, with six or seven brown spots......2. *albicapillus* Loew. ∨
3. Basal portion of the wings more or less brown, which color is gradually evanescent posteriorly:
   Frontal triangle (in the male) with a conspicuous covering of silvery white, shining, recumbent pile....3. *metopium* n. sp. ∨
   Frontal triangle (of the male) with blackish or golden-yellow, never conspicuous, pile:
   Face and cheeks with a dense beard of pale yellowish, shining; comparatively long hair, entirely concealing the color of the face under it ....4. *aurifer* n. sp.
   Face and cheeks with a sparse beard of mostly black pile, not dense enough to conceal the color of the face and cheeks under it; proboscis very long:
   Smaller species, with a fringe of golden pile round the mouth ....5. *cachinnans* n. sp.
   Larger species, with only black pile round the mouth,
   6. *lancifer* n. sp. ∨

1. BOMBYLIUS MAJOR Linn.—Dr. Loew identified with this European species specimens of the most common Californian *Bombylius*. I have about two dozen specimens, principally from Marin County, which vary in size from 7^mm to nearly 12^mm. The color is likewise variable in the more or less brown or yellowish shade of the fur, in the greater or lesser distinctness of the tufts of black pile on the sides of the abdomen, etc. In a number of specimens, there is no vestige of white fur on the chest and the mentum. Three specimens have the femora black and the tarsi dark brown. The females, of which I have five, show on the front part of the thorax the blackish spot, which distinguishes the southern variety of the European *B. major* (see Loew, Neue Beitr., iii, 14). Whether some of these varieties do not constitute different species, I do not pretend to decide. At the same time, I confess not to know in what the difference between *B. major* and *B. fratellus* of the Eastern States consists. Wiedemann (Auss. Zw., i, p. 583) merely mentions the absence of the tufts of black pile on the abdomen; but I have seen specimens with such tufts. Macquart (Dipt. Exot., ii, 1, 98; *B. vicinus*) says that *B. fratellus* (his *B. vicinus*) resembles *B. major*, but that its fur on the abdomen is fulvous instead of yellow; that is all. Loew (l. c.) adopts *fratellus* as a separate species, but does not throw any light on the subject of its difference from *B. major;* he merely observes that apparently several species occur under that name in collections. As long, therefore, as it is not settled whether *fratellus* on one side and the Californian *major* on the other represent

one or several species, it is useless to try to discover a difference between them. I will only mention that Californian specimens frequently occur which are larger and of a deeper brownish-fulvous than any I have seen from the Atlantic States.

2. BOMBYLIUS ALBICAPILLUS Loew, Centur., x, 42.—Not rare in Marin and Sonoma Counties in April and May; a specimen from Yosemite Valley (June 13) has a whitish instead of yellowish fur.

3. BOMBYLIUS METOPIUM n. sp., ♂.—Frontal triangle with a conspicuous covering of long, silvery-white, recumbent hair, entirely concealing the ground-color; face clothed with brownish-gray pollen, and sparsely beset with long brownish-black pile; a fringe of fulvous pile along the oral margin; long white pile on the under side of the head; occiput with pale yellowish pile; occipital triangle black, with a few black hairs; antennæ black, with black hairs on the first two joints; third joint but little expanded in its proximal part, about once and a half the length of the two first taken together; proboscis of moderate length, about twice as long as the head. The fur on the body is pale yellowish, with some whitish reflections above the root of the wings; a stripe of dark hairs between the latter and the shoulders; chest with white pile; on the abdomen, rows of black hairs are visible on the posterior margins of the segments; on both sides of the second and the following segments, they are more dense. Stem of halteres brownish; knob yellow. Femora black, beset with yellowish scales; tibiæ and tarsi brownish-red, darker toward the tips. Wings brown at base and along the anterior margin, including the two basal, the marginal, and the proximal half of the first submarginal cells; on the inner surface of these cells, however, the brown is more diluted; cross-veins at the base of the first and fourth posterior cells, as well as the bifurcation of the second and third veins, are clouded with dark brown; the cross-vein at the base of the second posterior cell is about as long as the small cross-vein. Length 8–9^{mm}.

Hab.—Lagunitas Creek, Marin County, California, April 19. Although I have but a single specimen, I do not hesitate to describe this species, easily recognizable by the silvery tuft on the front of the male.

4. BOMBYLIUS AURIFER n. sp., ♂ ♀.—*Male.*—Epistoma with a dense mystax of pale golden-yellow hair, covering the edge of the mouth, but not quite reaching the lower corner of the eye; some few black hairs in the upper part of this mystax near the orbit; front clothed with shorter hairs of the same color (but by far not as long and conspicuous as the snow-white pile on the front of *B. metopium*). Proboscis a little more than twice the length of the head. The two first antennal joints with black pile; the third about once and a quarter the length of the two first, rather broad, its greatest expansion beyond its middle; rather suddenly attenuated at the tip. Under side of the head with whitish pile; on the occiput, it is more yellowish-white. The fur of the body rather uniformly pale yellow; a small tuft of black pile on each

side of the third segment; the fur on the chest more whitish; ground-color of the thorax deep black. Halteres yellow. Legs black; base of tibiæ brown; femora and tibiæ beset with whitish-yellow scales. Prox-imal part of the wings tinged with pale reddish-brown as far as the tip of the first vein and the anterior and posterior cross-veins; the brown gradually evanescent about this limit; the remainder of the wing gray-ish-hyaline.

Female.—Like the male in all respects, except that the fur is much more whitish, including that of the mystax. The front is clothed with some scattered, erect, whitish pile, and a very dense, recumbent, pale whitish-yellow, shining tomentum, completely covering the ground-color, except on the vertical triangle and a line in the center of the front; some black hairs above the mystax, but none on the front and vertex; the brown on the wings less dark and extended; the ground-color of the femora and of a part of the tibiæ concealed under a thick covering of whitish scales.

Length about 6.5^{mm}.

Hab.—Webber Lake, Sierra Nevada, California, July 25. A male and a female. Will be easily recognized by the dense hairy clothing of the face and the shape of the antennæ. The paler fur of the female speci-men I hold to be varietal and not sexual.

5. BOMBYLIUS CACHINNANS n. sp., ♀ ♂.—Body black, densely clothed with a dull yellowish fur; epistoma brownish-yellow in the female, darker in the male, and covered with grayish pollen; the longer hairs upon it are black; the shorter pile round the edge of the mouth is golden-yellow. Frontal triangle in the male grayish-pollinose, and with black pile; in the female, the opaque grayish-black front is beset with a recumbent reddish-golden short tomentum, the ocellar tubercle and sur-roundings being free from it; some scattered black erect hairs are visible on the sides of the front and on the vertex; proboscis nearly as long as the body; antennæ with black pile on the basal joints; the third joint is rather narrow in the male and somewhat broader in the female, and of equal breadth for more than half of its length, beyond which it is narrower. The fur on thorax and abdomen is of a nearly uniform color; on the hind margins of the segments of the latter, some sparse black hairs may be perceived, which appear as indistinct tufts on the sides of the second and of the following segments. Wings gray-ish-hyaline, tinged with pale brown or reddish-brown at the base and in the costal cell; in the female, the brown does not fill out the distal half of the first basal nor the second basal cell; in the male, the brown is darker, and gradually evanescent posteriorly, but it extends over nearly the whole wing. Legs yellowish-red; tarsi, except their base, black; in the male, the base of the femora is black; the knees have black dots on the front side. Length of male 6.5^{mm}; of female 7–8^{mm}.

Hab.—Sonoma County, California, April 27–May 9. Two females and one male.

6. BOMBYLIUS LANCIFER n. sp., ♂ ♀.—Body black, densely clothed with yellow fur; tufts of brown pile in the posterior corners of the thorax, and two tufts of black pile on each side of the abdomen connected by rows of black pile over the back; a stripe of dark brown pile between the humerus and the root of the wing; on the chest, the hair is paler; on the mentum, white. Epistoma yellowish-brown, shining above, grayish-pollinose on the sides, beset with black pile; frontal triangle in the male grayish-pollinose, beset with black pile; the whole front and vertex in the female grayish-pollinose, beset with some recumbent golden-yellow tomentum and longer black pile. Proboscis long, as long as the body, perhaps a little longer. Third joint of the antennæ moderately broad, with parallel sides, on its last third tapering toward the tip. Legs red, thinly clothed with whitish scales, and beset with black spines; tarsi brown, reddish at base; knees dark brown, especially on their anterior side. Wings blackish-brown on their basal half; strongly tinged with grayish on the rest of the surface. Length 10^{mm}.

Hab.—San Francisco, Cal. (H. Edwards); Yosemite Valley (June 9). One male and two females. The fur is intact in the male only; that of the female, which I took in Yosemite Valley, is more whitish. In general appearance, *B. lancifer* is not unlike *B. varius* of the Atlantic States; but the latter has a much shorter proboscis, a distinctly lanceolate third antennal joint, black pile on the chest, wings less grayish on their distal portion, etc.

ANASTŒCHUS nov. gen.

Closely allied to *Systœchus*, but easily distinguished from the North American species of that genus by the following characters:—

Head comparatively larger, and front of the female broader.

Face, cheeks, and *lower part of the front* are beset with erect pile, which forms a dense broad brush, entirely concealing from view the outlines of the mouth and cheeks, as well as the basal joints of the antennæ. When the pile is removed, the face shows a structure entirely different from that of *Systœchus;* in the profile, the mouth, instead of projecting forward, has its sides, the cheeks, on the same plane with the eyes, and even somewhat withdrawn behind them; the epistoma, or face above the mouth, projects very little, and descends almost directly below the antennæ.

Eyes, in the male, separated by an interval on the vertex, which is not coarctate in front of the ocelli; a distinct oblique line separates the upper and larger facets from the lower and smaller ones (in *Systœchus,* the narrow interval between the eyes on the vertex is strongly coarctate in front of the ocelli; the passage between the two kinds of facets is gradual and imperceptible); in the female, the interval between the eyes is about one-half broader than the horizontal diameter of the eye.

Antennæ of the same structure as in *Systœchus,* but the third joint, beyond the usual ring-like expansion at the extreme base, is, for a cer-

5 H B

taiu distance, more distinctly attenuated, the dilatation being removed farther toward the middle; the slender distal half is more elongated.

Venation like that of *Systœchus*, but the relation between the cross-vein at the base of the second posterior cell to that at the base of the third is like 1 to 2 or 3 here, while it is like 1 to 5 or 6 in *Systœchus;* this causes the proximal end of the third posterior cell to be less long and pointed in *Anastœchus*, and renders the discal cell a little broader; the usual enlargement of the costa at the base, besides the usual pile, bears a fringe of bristles.

General outline of the body more elongated than in *Systœchus;* the hairs on the end of the body longer, tuft-like.

This genus answers Dr. Loew's first and smaller division of *Systœchus* (Neue Beitr., iii, 35). The structural differences are of an importance which not only justifies but requires the formation of a new genus.

Anastœchus means "separated", in contradistinction from *Systœchus*, "belonging together".

ANASTŒCHUS BARBATUS n. sp., ♂ ♀.—Densely clothed with grayish-yellow pile, mixed with some black pile at the end of the abdomen (especially in the male); beard white, with some black hairs above; wings grayish-hyaline, more or less brownish at the base. Length 5–7^{mm} (exclusive of the length of the pile).

Ground-color of the body grayish-black, densely clothed with long pale grayish-yellow pile (much less yellow than that of *Systœchus vulgaris*). Head with a dense beard of white pile, slightly yellowish round the base of the antennæ; above it, on the front, a tuft of long black hairs, descending on each side along the eyes to about the middle of the inner orbit in the male, much less in the female. Antennæ black. Chest with white pile. On the abdomen, besides the prevailing yellowish pile, darker hairs are visible in rows, on the posterior margins of the segments; they are black, and especially visible on the last two or three segments, so that in most of the male specimens they impart a blackish tinge to the pile around the tip of the abdomen; the extent and number of these black hairs is, however, very variable, and in most female specimens they disappear altogether. Knob of halteres yellow. Femora, except the tips, black, but densely clothed with white scales; tip of femora, tibiæ, and tarsi reddish-yellow; the usual spines of the same color; end of tarsi brown; in the female, the yellow on the femora is more extended. Wings grayish-hyaline, with a more or less extended shade of brown at base, which is almost obsolete in the female; costa at the root with short yellowish-white and longer black pile, the latter forming a kind of comb; wing-veins black, those at the root and near the costa often pale brownish.

Hab.—Cheyenne, Wyo., where I found it commonly on the 21st of August, 1876. Five males and as many females; the latter smaller. Besides these, I have three specimens from the Twin Lakes, Colorado (9,300 feet altitude, collected by Lieut. W. L. Carpenter), which I cannot distinguish from the others, although they measure 10^{mm}.

Two male specimens from California (one of them from the Shasta district, H. Edwards) belong apparently to the same species, and are nearly of the size of the larger ones from Cheyenne.

Finally, a female from Nantucket, Mass. (caught over sandy soil, middle of September, by Mr. S. H. Scudder), is larger than the females from Cheyenne, measuring nearly 10^{mm}; the proboscis is a little shorter and there is more yellowish pile in the beard around the antennæ; in other respects, the agreement is perfect.

A. barbatus is therefore either a species with a very wide distribution, or else there are several closely allied species, which, with the material before me, I am unable to distinguish. The European species of this group (*A. nitidulus*, etc.) are also remarkably like *A. barbatus*.

SYSTŒCHUS.

A difficult genus on account of the great resemblance of the species and the apparent scarcity of distinctive characters of an absolute and more than comparative value. *Systœchus* is very common in the West, but occurs also in the Southern States. I have seen only one specimen from the Northeastern and Middle States, which I took near Alexandria Bay, Saint Lawrence River.

1. SYSTŒCHUS VULGARIS Loew, Centur., iv, 52 (Nebraska; Dr. Hayden). A common species in Colorado, about Denver, Manitou, etc., July, August (P. R. Uhler and myself); also in Utah (a male from Salt Lake City, July 21, by A. S. Packard). The fulvous hairs on the face are often more abundant than Dr. Loew's wording implies; in the female, they extend to the front, especially along the eyes.

A female specimen which I caught near Alexandria Bay, N. Y., on Saint Lawrence River, is smaller, but does not, in other respects, differ from *S. vulgaris*.

I have two males and two females from Illinois (Le Baron) and Dennison, Crawford County, Iowa (Allen; July, 1867), which, instead of the usual pale yellow color of *S. vulgaris*, are of a decidedly reddish-yellow, almost rufous; in outline, they seem to me broader than *S. vulgaris*, and may belong to a different species.

2. SYSTŒCHUS SOLITUS Walker, List, etc., ii, 288 (*Bombylius*).—Florida. As suspected by Dr. Loew (Centur., iv, 52), this is a *Systœchus*. I have a specimen from Capron, Fla. (Messrs. Hubbard and Schwarz, in April), which answers the description. It differs from *S. vulgaris* in having the tarsi and the ends of front and hind tibiæ black; the spines on femora and tibiæ are of the same color. I do not discover any other difference.

3. SYSTŒCHUS CANDIDULUS Loew, Centur., iv, 51 (Wisconsin).—Besides the whitish pile covering the whole body, this species is easily distinguished by its longer proboscis and more hyaline wings, with paler veins; the pile on the face and front is black without admixture, and

only a slight grayish or whitish pollen is visible under it on the face. I have specimens from Illinois and Kansas.

4. SYSTŒCHUS OREAS n. sp.—Differs from *S. vulgaris* in the third antennal joint being a little broader, the mystax more mixed with fulvous pile, the proboscis longer, the legs darker, the wings more grayish, the covering of pile more dense and of a paler shade of yellow, the ground-color less dark (when denuded), without reddish on the scutellum; on the average, the size is somewhat larger.

Male.—The blackish-gray ground-color of the body is entirely concealed (in intact specimens) under a thick covering of pale yellow pile, giving the body an elongated-oval shape, slightly broader about the middle of the abdomen; face and front clothed with a recumbent fulvous tomentum and erect black pile; mystax mixed of both; some black pile on the vertex; antennæ black, third joint considerably expanded on its proximal half; legs black; femora densely covered with the usual appressed whitish hairs, which conceal the ground-color; tibiæ reddish, but clothed with the same whitish pubescence; the latter part of the tibiæ is black, and on the inner side this color extends farther upward than externally; tarsi deep black. Wings with a decidedly grayish tinge, brownish-yellow at the base and in the costal and first basal cells. Length about 10^{mm} (including the length of the pile at both ends of the body, but excluding the antennæ).

Female.—I have a single somewhat damaged specimen, which evidently belongs here, although it is smaller, and the femora and tibiæ, except the tip, are yellowish-red. Length about 8^{mm}.

Hab.—Webber Lake, Sierra County, California, July 22–26. Three males and one female. None of my specimens show any reddish on the scutellum.

PANTARBES nov. gen.

Belongs to the *Bombylina*, with a closed first posterior cell, but differs abundantly from *Bombylius*, *Systœchus*, and *Anastœchus* in having three submarginal cells; the front very broad in both sexes; the antennæ remarkably distant at base, and with a much more developed, 2-jointed, terminal style; the ends of the second vein and of the anterior branch of the third strongly curved and bent forward (as in *Ploas* and *Lordotus*).

In the thickness of its beard, entirely concealing the outlines of the mouth, it resembles *Anastœchus*, but it surpasses it in the breadth of the head; its mouth is much smaller. The proboscis is shorter here than in any of the above-named genera, and not attenuated toward the tip.

Its closest relative, however, is perhaps *Mulio* (as understood by Meigen, Eur. Zweifl., ii, tab. xvii, f. 26–28), with which it shares the shape of the head, the distant eyes in both sexes, the distant antennæ, and the general appearance of the body. But *Mulio* has the first posterior cell open, no pulvilli, the first antennal joint much shorter, for an antennal style a mere bristle, the beard less long, etc. Comparatively, the head of *Pantarbes* is larger and broader, the body is stouter, the præfurca shorter.

Head large, considerably broader than the thorax; vertex in the male equal in breadth to about two-thirds of the greatest horizontal diameter of the eye, still broader in the female; the front immediately above the antennæ is about three times as broad as the vertex (in the male), the eyes being placed obliquely; front and face descend nearly perpendicularly toward the oral margin, the antennæ, therefore, inserted at a much lower level than the vertex. The lower part of the front, the face, the base of the antennæ, and the oral opening are entirely concealed from view by a dense tuft-like crop of hair, occupying the whole anterior part of the head, and similar to that of *Anastœchus*. Those parts of the head can only be examined after the removal of this hair.

Ocelli placed on a hardly perceptible flat prominence of the vertex; the lateral ones large, and at a distance from each other, which, in the male, is at least by one-half larger than the interval between each of them and the nearest orbit of the eye; in the female, the latter interval is a little larger than the distance between the ocelli.

Antennæ inserted at a distance from each other which is but little shorter than the breadth of the vertex in the male; first joint (concealed in the facial tuft of hair) nearly cylindrical; second joint short, not longer than broad; the third a little longer than the two first taken together, slender, beginning by a short basal expansion, then attenuated for about one-third of its length, and then again very slightly expanded, with but a small attenuation toward the end; at the tip, a minute, stout, cylindrical, 2-jointed style, with a microscopic bristle on top; the style when viewed from above the vertex is somewhat at an angle to the rest of the antenna.

Eyes glabrous; the passage from the larger facets above to the smaller ones below in the male is gradual; in dry specimens, at least, the line of separation is not visible.

Oral opening comparatively small, its upper edge reaching but little above the lower corners of the eyes.

Proboscis porrected forward, comparatively short, projecting but little beyond the tip of the antennæ, not tapering toward the end.

Thorax of moderate size, not gibbose, nearly on a level with the head.

Scutellum small in comparison with those of *Bombylius* and *Systœchus*.

Abdomen a little longer than thorax and scutellum together, as broad as the thorax at the base, and gradually tapering toward the tip.

Legs, especially the femora, comparatively strong; first tarsal joint a little shorter than the four others taken together; *ungues* curved; *pulvilli* distinct and long.

Venation of the wings : first posterior cell closed; its petiole as long as in an ordinary *Bombylius;* second vein gently arched before the cross-vein, connecting it with the third (but not as strongly curved as in *Lordotus*); beyond this cross-vein, its curvature is stronger than in *Lordotus*, so that the expanded distal end of the marginal cell bulges out beyond the end of the first submarginal cell; three submarginal cells formed by a cross-vein connecting the second vein with the anterior branch of the third very near its base; the first of the two exterior submarginal cells almost crescent-shaped, in consequence of the curvature of the veins forming it; small cross-vein about the middle of the discal cell, and hence the first basal cell much larger than the second; the bifurcation of the second and third vein takes place a little before the middle of the distance between their common root and the small cross-vein; these two veins become at once distinctly divaricate (and not approximate and parallel for a considerable distance, as in *Bombylius* and *Systœchus*); the rest of the venation as in those two genera,— that is, anal cell open, etc.

Pantarbes, in Greek, means *full of fear*.

PANTARBES CAPITO n. sp., ♂ ♀.—Body grayish-black, densely clothed with whitish-gray pile; beard white; wings grayish-hyaline, the anterior half for about three-quarters of the length infuscated. Length 6–10^{mm}.

Front and lower part of the head and occiput densely clothed with snow-white pile; upper part of front with a fringe of long black hairs, which extend some distance downward along the orbits of the eyes; vertex likewise with a bunch of black hairs. Antennæ: first and second joints yellowish; the third black. Halteres yellow. Femora black, densely beset with white scale-like hairs, and some longer pile; tibiæ and tarsi reddish, the latter black toward the tip. The brown color of the wings extends from the root to the end of the first longitudinal vein, and a little beyond the small cross-vein; it gradually fades away posteriorly; anal and axillary cells hyaline.

Hab.—Sonoma County, California, April 27 to May 9; not rare. Nine males and one female. The latter is, of all the specimens, the smallest; its wings are less infuscated at the base; the beard round the antennæ is somewhat yellowish. In flying, this species frequently alighted on the soil.

COMASTES nov. gen.

Venation, antennæ, and proboscis of a *Bombylius*, but general outline of the body and the character of the fur of pile upon it entirely different. Head larger; thorax much longer; abdomen, on the contrary, smaller; the outline of the body more parallel, less ovate; scutellum much larger; hind legs longer. The hair on the epistoma is less long and bushy, more recumbent, which gives the large, broad head, especially when seen from above, a totally different appearance. The fur on the thorax is dense, but shorter than in *Bombylius*, more like that of an *Eristalis*; that on the abdomen is as long, but less erect and less evenly distributed than in *Bombylius*.

I have only a single specimen, which, from the breadth of the front, I judge to be a female.

Head transverse, inserted as in *Bombylius*, a little lower than the thorax; as broad as the latter in its broadest part (even a little broader, if the fur be removed); the interval between the eyes, in the female, broad, very little narrower on the vertex than near the mouth; three large ocelli on a flat protuberance; eyes reniform; mouth oval, rather large; epistoma moderately projecting in the profile in front of the eyes; cheeks not projecting and head not descending below the eyes; occiput but little swollen, densely clothed with down.

Proboscis long, three-quarters of the length of the body, and stouter than in a *Bombylius* of equal size; palpi elongated, second joint short.

Antennæ approximated at base; first joint elongated, cylindrical; second not much longer than broad; the third one-third longer than the two first taken together, narrow, linear on its first half, gradually tapering on the second, truncate at the tip, upon which is inserted a short 2-jointed style. I do not perceive any terminal bristle in my specimen. The whole length of the antenna is about equal to the distance between the ocelli and the mouth.

Thorax rather long, square, with nearly parallel sides, moderately convex, densely clothed with short erect pile above and with longer hairs on the pleuræ; scutellum comparatively larger than in *Bombylius*, almost semicircular, moderately convex.

Abdomen short, much smaller in bulk than the thorax, turned down at the end, unevenly clothed with long pile arranged in semi-erect rows and tufts, which begin at the posterior end of the second segment; venter hollow.

Legs like those of *Bombylius*, only the hind pair comparatively longer; pulvilli much shorter than the ungues.

Wings and *venation* as in *Bombylius;* the contact of the second submarginal cell with the first posterior is very short, almost punctiform; the same is the case with the second posterior and the discal cell; small cross-vein about the middle of the discal.

Comastes, in Greek, means *a reveler.* (I am aware of the existence of *Comaster* Agassiz, *Echinod.*, but both the termination and the derivation of that word are different.)

COMASTES ROBUSTUS n. sp., ♀.—Ground-color of the head grayish-white, yellowish round the mouth, densely clothed with pale yellowish-white pile, more yellow on the front; a tuft of black pile on the ocellar tubercle. Antennæ black; first joint grayish-pruinose. Palpi reddish, second joint brownish. Proboscis black. The dense pile on the vertex is yellowish above, whitish below. The grayish-black ground-color of the thorax is almost concealed on the dorsum by a dense, short, erect clothing of fulvous pile; on the pleuræ, a tuft of whitish-yellow hair; that on the chest almost white. Scutellum reddish, with fulvous pile and some black bristles along the edge. Abdomen blackish-gray; second segment with a short, appressed tomentum, forming a yellowish-white cross-band; the remaining segments, beginning with the posterior margin of the second, are covered with long, semi-erect, black pile, across which, in the middle of the abdomen, there is a triangular figure formed by similar pile, but white; the apex of the white triangle rests on the hind margin of the second segment; the oblique stripes of white pile, forming the sides of the triangle, run downward toward the venter; the inner side of the triangle is filled, partly with black, partly with white pile, the latter chiefly occupying the end of the abdomen; the venter is clothed anteriorly with white, posteriorly with black pile. Legs reddish; tarsi darker; hind tibiæ and tarsi reddish-brown. Knob of halteres yellow. Wings grayish-hyaline; base as far as the basal cross-veins brownish; costal cell pale yellowish. Length 11–12^{mm}.

Hab.—Waco, Texas (Belfrage). A single female. The specimen is in Mr. E. Burgess's collection in Boston.

LORDOTUS.

LORDOTUS GIBBUS Loew, Centur., iv, 53.—Dr. Loew described a female from Matamoras. I have a dozen specimens from Denver, Colo. (Uhler, August 18), Cheyenne, Wyo. (myself, August 21), and California (San Francisco and Shasta district, H. Edwards). The color of the antennæ is variable. In all the specimens from California (six females), the two basal joints are red. One of the specimens from Denver has the second joint red toward the tip only, as described by Dr. Loew. In the other specimens from Denver, and also in those which I took near Cheyenne, the antennæ are altogether black, although the basal joints are grayish-pollinose. Well preserved specimens show two grayish stripes on the thorax. The brownish-red color at the base of the wings and in the costal cell is often extended to the first basal and submarginal cells. A gray cloud is often visible on the cross-vein at the base of the fourth posterior cell. The costa in all my specimens is reddish, and not black, as described by Dr. Loew.

I have only a single male, taken near Cheyenne, Wyo. The eyes are closely contiguous on a rather long line, down to very near the base of the antennæ. The difference between the upper larger facets and the lower smaller ones is well marked, although the line of division between them is not very sharp (in the female, the facets are uniform); frontal triangle very small, glabrous. On the last abdominal segments, beginning with the fifth, many black hairs are mixed with the yellow ones, especially on the sides. The femora, except the last quarter, are black; the tarsi altogether black. The body is smaller and much more slender than that of the female.

As one of the specimens from Denver has been communicated to Dr. Loew, there can be no doubt about the specific identity.

LORDOTUS(?) PLANUS n. sp.—I place provisionally in this genus a Californian species, of which I have only a single male specimen, and for which the erection of a new genus will perhaps be necessary. It has the characters of *Lordotus*, except the general shape of the body, which is much less gibbose. The venation is exactly like that of *Lordotus*, including the remarkable sweep of the second vein. The second joint of the antennæ is comparatively shorter, as it is but little longer than broad.

Male.—Thorax clothed with yellowish-gray, abdomen with whitish pile; legs and antennæ black, the former densely clothed with an appressed white scale-like tomentum. Length 7–8mm.

Antennæ black, the first two joints beset with black pile, especially long on the under side; cheeks and face with grayish pile, with an admixture of black hairs in the mystax; the small frontal triangle clothed with whitish pollen; occiput with pale yellowish-gray pile. The grayish-black ground-color of the body is concealed under a dense covering of dull yellowish-gray pile on the thorax and of white pile on the abdomen. Knob of halteres yellowish-white. Wings subhyaline; veins brown, those nearer to the base and to the costa yellowish-brown; a darker spot on the first vein, at the junction of the cross-vein at the proximal end of the first basal cell; a similar spot, with a vestige of a cloud, on the præfurca; vestiges of clouds on the large and small cross-veins.

Hab.—Marin County, California (H. Edwards). A single male.

SPARNOPOLIUS.

1. SPARNOPOLIUS COLORADENSIS Grote, Proc. Entom. Soc. Phil., vi, p. 445.—Mr. Grote describes the female; the male stands in the same relation to it as the male of *S. fulvus* to its female; it is more slender in shape, and paler yellow; less fulvous in the coloring of its pile; the hairs on the antennal scapus are black. In the female, those hairs are variable in color, in some specimens black, in others mixed with bright fulvous ones; in others again the fulvous pile prevails. I have a number of specimens collected about Colorado Springs by Mr. Uhler.

2. SPARNOPOLIUS BREVICORNIS Loew, Centur., x, 43.—Waco, Texas; female. I have specimens of both sexes from the same locality. This species is exceedingly like the preceding; antennæ and proboscis, in my specimens at least, are shorter; the costal cell a little more yellowish; the fur a little less dense, especially in the female. I do not perceive any other differences.

3. SPARNOPOLIUS CUMATILIS Grote, Proc. Entom. Soc. Phil., vi, p. 445.—Colorado; female. I have never seen this species.

4. SPARNOPOLIUS FULVUS Wied., i, 347 (syn. *Bombylius L'herminieri* Macq., D. E., ii, 1, 103; *Bombylius brevivostris* Macq., l. c.).—A well-known species from the Atlantic States.

PLOAS.

The Californian species which I have may be tabulated thus:

Halteres with a yellow knob:

Wing-veins clouded with dark brown, but the inside of most of the
cells hyaline... ....1. *fenestrata* n. sp.

Wings brown at base, which color gradually fades into grayish
posteriorly:

Thorax and proximal half of the abdomen clothed with yellowish-rufous pile above and below......2. *rufula* n. sp.

Whole body clothed with black pile, mixed with yellowish-
gray:

Large species.........................3. *nigripennis* Lw.

Small species...........................4. *atratula* Lw.

Halteres with a brown knob:

Abdomen metallic bluish-green:

Abdomen opaque anteriorly and in the middle, its sides and
apex being shining bluish-green; pile on thorax and on
the upper side of abdomen rufous.......5. *obesula* Lw.

Abdomen shining greenish-blue; pile on thorax and proximal
half of the abdomen above pale yellow; on the distal
half and below black.................6. *amabilis* n. sp.

Abdomen black..................................7. n. sp. indescr.

None of the species described below has any marked impression on
the scutellum, and all have three submarginal cells.

Ploas limbata Loew, Centur., viii, 51, from New Mexico, I do not know.

1. PLOAS FENESTRATA n. sp., ♂ ♀.—Wings dark brown along the anterior margin; all the veins (except the seventh) and cross-veins broadly
clouded with brown. Hyaline spaces in the following cells: the two
outer submarginal and the distal half of the inner submarginal; the end
of the first posterior; the three other posterior; the discal; the whole axillary; nearly the whole anal (except at both ends). A hyaline spot in the
distal half of the second basal cell. Head and thorax beset with long
black and shorter pale yellowish-gray pile, the latter thicker and longer
on the occiput, the chest, and the pleuræ. Antennæ black, with a very
stout first joint and an unusually elongated subcylindrical second joint
more than half as long as the first. Thoracic dorsum black, opaque;
scutellum, when denuded, shining. Abdomen black, opaque; on the
hind margin of each segment, a cross-band of whitish-gray recumbent
pubescence, which expands in the middle so as to coalesce with the preceding cross-band; in the middle of each cross-band, on the hind margin of each segment, there is a more or less triangular spot of brownish-
fulvous hair; these spots gradually diminish on each subsequent segment; the whole abdomen is beset, besides, with black, erect pile.
Legs black in intact specimens, with a dense covering of brownish-ful-

vous scales, more whitish on the anterior part of the femora. Halteres reddish-yellow. Length 10–11^{mm}.

Hab.—Crafton, near San Bernardino, Cal., in March; San Rafael, Cal., and Sonoma County, in April and May. Much less common than *P. nigripennis.* Three males and one female.

- This species, in several respects, is peculiar. The marginal cell is unusually short, reaching very little beyond the tip of the first vein, and not expanded at the end, as in all the other species. The consequence is that the submarginal cells have a shape different from the usual one : the first outer submarginal cell is larger, the inner submarginal broader toward the end. The structure of the antennæ is peculiar in the shape of the second joint, which is about as long as the third. The rather slender abdomen has none of the dense fringe of pile along the lateral edges, which distinguishes most of the following species.

2. PLOAS RUFULA n. sp., ♂.—Second antennal joint less than half the length of the first; two basal joints beset with long black pile; the under side of the head, occiput, thorax above and on the sides, and the abdomen beset with rufous pile; conspicuous tufts of black pile on the sides of the three penultimate abdominal segments; ground-color of the abdomen black, opaque on the two first segments; the other segments are shining greenish-black, with a small black opaque triangle in the middle and a narrow opaque cross-band at the base. Halteres reddish-yellow; legs black, beset with fulvous scales and pile, principally on the femora; spines on the tibiæ black; wings grayish, infuscated at base and along the fore border as far as the end of the first vein and including the first basal cell; the brown gradually evanescent; small cross-vein with a deep brown cloud; posterior cross vein with a weaker one. Length 11–12^{mm}.

Hab.—San Geronimo, Marin County, Cal., April 19. Two males. In one of them, the cross-vein separating the second outer submarginal cell from the inner one is wanting on both wings.

3. PLOAS NIGRIPENNIS Loew, Centur., x, 45.—This is the most common Californian *Ploas.* I have numerous specimens of both sexes from Crafton, near San Bernardino, in March; Marin and Sonoma Counties, in April and May; Yosemite Valley, in June; Webber Lake, Sierra Nevada, in July. Dr. Loew describes the female. In the male, the opaque spaces in the middle of the third and following abdominal segments are much broader.

4. PLOAS ATRATULA Loew, Centur., x, 44.—I refer to this species, with a doubt, two female specimens taken by me near the Geysers, Sonoma County, California, in May.

5. PLOAS OBESULA Loew, Centur., x, 46.—California. Two male specimens, received from Mr. H. Edwards, without indication of the precise locality, I unhesitatingly refer to this species.

6. PLOAS AMABILIS n. sp., ♂ ♀.—Head black, beset with black pile, except on the occiput, where it is yellow; the front of the female, be-

sides the long black pile, in some specimens shows some shorter, yellow hairs. The whole under side of the body as well as the three last abdominal segments above are beset with deep black pile, which forms a dense fringe on the edge of those segments, especially long, and consisting of a row of tufts in the male. The upper side of the thorax, as well as the whole anterior half of the dorsal side of the abdomen, is clothed with pale yellow (straw-colored) pile. The ground-color of the thorax is black; that of the abdomen greenish-blue, shining, except the first segment, which is opaque, blackish; wings grayish-hyaline, brown at base and along the anterior margin as far as the end of the first vein, the brown gradually evanescent posteriorly; halteres black; leg black. Length 9-10^{mm}.

Hab.—Yosemite Valley, California, where I caught two females and one male of this pretty species, June 5-15.

7. PLOAS n. sp., ♀.—I have a single specimen from Yosemite Valley, June 14, measuring 5-6^{mm} without the antennæ; knob of halteres black, except the base, which, like the stem, is yellow; first joint of the antennæ unusually long and stout, the second cylindrical, about one-third as long as the first. The specimen being denuded is unfit for a description. The body is uniformly black; tufts of yellow pile are left on the sides of the thorax. I mention this species in order to call the attention of collectors to it.

PARACOSMUS

(nomen novum, vice *Allocotus* Loew).

PARACOSMUS EDWARDSI Loew, Centur., x, 48 (*Allocotus*).—The name given by Mr. Loew to this new genus being pre-occupied (*Allocotus* Mayr, Hemipt., 1864; *Allocota* Motchulski, Coleopt., 1859), I have changed it to *Paracosmus* (meaning, in Greek, *disorderly*).

Loew describes the female. In the male, the eyes are not contiguous on the front, which is but little narrower than that of the female. The eyes have uniform facets above and below. The hypopygium is rather large for the family, consisting of subhemispherical lower piece and a forceps-like organ above, with broad valves.

I found this curious insect in both sexes, flying in the sun over the sands round Lone Mountain, San Francisco, June 29.

PHTHIRIA.

I do not possess *P. punctipennis* Walker (List, iii, 294) from Georgia. *P. sulphurea* seems to have a wide distribution, from New Jersey to Colorado and Texas. The other species seem to be more exclusively western or Californian.

1. PHTHIRIA SULPHUREA Loew, Centur., iii, 18 (New Jersey, female).— I have specimens from Waco, Texas (Belfrage, communicated by Mr. Burgess), Colorado Springs (Uhler), and Illinois (Le Baron), which ap-

parently belong here. The antennæ are yellow, therefore paler than the description makes them; the sheath of the proboscis is variable in its coloring, being sometimes entirely black, sometimes yellow, except the lips, which remain black; the costa is yellowish-brown (and not black, as the description has it). The specimens from Illinois are smaller, and have no stump of a vein in the discal cell. These discrepancies notwithstanding, I could not take my specimens for a different species before comparing them with the types of the description. *P. sulphurea* is figured in Mr. Glover's Manuscript Notes, etc. (Diptera, tab. v, f. 1). The male of this species has the abdominal segments tinged with brownish at the base, the hind margins remaining sulphur-yellow.

2. PHTHIRIA SCOLOPAX n. sp.—Drab-colored; thorax with faint yellow lines; legs yellow, tips of tarsi black; wings large, all the cross-veins strongly, all the ends of longitudinal veins and the distal half of the costa more faintly, clouded with brown; proximal ends of the second submarginal and of the third posterior cells square, and both provided on the outside with a long stump of a vein. Length 6–7^{mm} (without the proboscis).

Head yellowish; cheeks, except the orbits, dark brown or reddish-brown, shining, with a yellow cross-line in the middle; ocellar triangle in the male dark brown, grayish-pruinose; frontal triangle yellow in the middle, reddish-brown on the sides, which color is separated by a yellow line from the brown of the cheeks; in the female, the interval between the eyes is yellowish-brown, with a dark brown spot on the vertex, upon which are the ocelli; the orbits of the eyes are sulphur-yellow. Palpi long and slender, dark brown. Antennæ yellowish-brown, last joint more brown, nealy three times the length of the two first taken together, its sides nearly parallel, its tip distinctly emarginate. Proboscis nearly as long as the body in the male, somewhat shorter in the female. Thorax opaque yellowish-gray, beset with an appressed golden tomentum; on the dorsum, two pale yellow longitudinal lines, and a third, much more delicate one, between; lateral margins of the dorsum and antescutellar callosities likewise yellowish; scutellum with a yellow line in the middle, and often with a brown spot on the tip. Halteres yellow, with a brown spot on the knob. Abdomen brownish-yellow. Legs yellow, tarsi, except the base, dark brown. Wings rather large and broad; the proximal ends of the second submarginal and of the third posterior cells are square; each emits on the outside a long stump of a vein, projecting, the one into the first submarginal, the other into the discal cells; the cross-veins at the base of the second submarginal and of the four posterior cells are clouded with dark brown, which clouds extend along the above-mentioned stumps; a large cloud at the bifurcation of the second and third veins; the costal margin, especially beyond the end of the auxiliary vein, and the ends of all the longitudinal veins, are also clouded. In most specimens, there is a curved stump of a vein with a cloud upon it, on the second vein, opposite the small cross-vein, and inside of the first submarginal cell.

Hab.—Manitou, Colo., August 18.   One male and three females. A note which I took when the specimens were alive describes the eyes as greenish-purple, with a bluish-purple cross-band across the middle; in the male, the facets *above* the cross-band are the larger; in the female, those *below*.

P. scolopax is very like the figure of *Poecilognathus thlipsomyzoides* Jaennicke (Neue exotische Diptern, tab. i, f. 11); and although it appears from the description that it is a different species, it is equally evident that both are most closely related.   I do not see any sufficient ground for separating this species from *Phthiria;* at any rate, it is singular that Mr. Jaennicke, in his definition of his new genus *Poecilognathus,* does not even mention *Phthiria.*   He says the venation is that of *Thlipsomyza;* but that genus has the first posterior cell closed, while *Poecilognathus,* according to the figure, has it open.

3. PHTHIRIA EGERMINANS Loew, Centur., x, 47.—California.

4. PHTHIRIA NOTATA Loew, Centur., iii, 19.—California.

I have not seen these two species.

5. PHTHIRIA HUMILIS n. sp., ♂.—Grayish-black; sparsely beset with pale grayish-yellow pile on the thorax, and with white pile on the abdomen and the under side of the head; legs and antennæ black; wings hyaline.   Length 4–5ᵐᵐ.

The grayish-black ground-color of the epistoma, cheeks, and front is clothed with a grayish pollen, more yellowish-gray on the occiput; oral margin and chin beset with white pile, silvery in a certain light; occiput with yellowish pile; antennæ black.   The grayish-black ground-color of the thorax is but very little concealed above by a covering of pale grayish-yellow downy pile, more dense anteriorly; pleuræ opaque, almost glabrous; abdomen sparsely beset with long, erect, whitish pile, more dense on the sides; halteres brownish-yellow; legs black; femora with some whitish pile; wings hyaline; posterior costal cell (interval between the auxiliary and first vein) yellowish in its latter half; veins dark brown, except near the root.

Hab.—Los Guilucos, Sonoma County, Cal., July 4.   A single male.

In life, the smaller facets on the lower part of the eye were of a darker color than those above.

This species can be, without difficulty, referred to *Phthiria,* although in its coloring and its pubescent body it differs from the other American species of the genus.   The flattened antennæ, truncate at the end, the long proboscis, and the venation, are those of *Phthiria.*   But the second submarginal cell and third posterior bear no stumps of veins on their outside, as they do in *P. sulphurea* and *scolopax.*

GERON.

The described North American species are :—

holosericeus Walker, List, ii, 295.—Georgia.

senilis Fab., Wied., i, 357.—West Indies (Wied.); Texas (Macq.).

calvus Loew, Centur., iv, 54.—New York.
macropterus Loew, Centur., ix, 76.—New York.
subauratus Loew, Centur., iv, 55; also ix, 77, nota.—Pennsylvania.
vitripennis Loew, Centur., ix, 77.—Middle States.
albidipennis Loew, Centur., ix, 78.—California.

SYSTROPUS.

Only a single species has been discovered in the United States, *S. macer* Loew (Cent., iv, 56). It occurs in all the Atlantic States. I have seen it from Kansas. I do not know whether it goes farther west or not.

S. macer has been bred from the cocoon of a *Limacodes*, the larvæ of which are allied to those of *L. pithecium* (see Walsh, in the Proc. Bost. Soc. N. H., vol. ix, 300, Febr., 1864). The fly, kindly communicated to me by Mr. Walsh after the publication of the article, is not a *Conops*, as he thought at the time, but *Systropus macer*. Quite recently, Mr. Westwood bred a species of *Systropus* from a South African cocoon, resembling that of *Limacodes* (Trans. Entom. Soc. London, 1876, 575).

LEPIDOPHORA.

LEPIDOPHORA ÆGERÜFORMIS Westw.—Occurs from Georgia to Kansas.

LEPIDOPHORA APPENDICULATA Macquart, suppl., i, 118.—Texas.

A third species, *L. (Toxophora) lepidocera* Wied., without locality, is mentioned by Macquart, l. c., as possibly the female of his species.

TOXOPHORA.

Two North American *Toxophoræ* have been described, *T. amphitea* Walk. and *T. leucopyga* Wied.; two have been figured, but not described, *T. fulva* Gray and *T. americana* Guérin.

T. leucopyga Wied., i, 361 (without *patria*), was referred by Macquart (ii, 1, 117) to a species from the Carolinas; this species has only two submarginal cells, and no stump of a vein in the second posterior; the third vein (and not the second) is furcate; both Wiedemann's (tab. v, f. 3) and Macquart's (l. c., tab. xiii, f. 1) figures agree in this.

T. americana Guérin is not described; the figure shows four complete posterior cells, and an abdomen with interrupted cross-bands, but no longitudinal stripe, as in both species described below.

T. fulva Gray, in Griffith's Anim. Kingd., Insects, tab. 126, f. 5, from Georgia (Walker, List, etc., ii, 298), is described (l. c., 779) thus: "fulvous, with a black mark on the thorax and black lines across the abdomen". The figure agrees with this statement (it can hardly be called a description).

The metamorphosis of *Toxophora* was hitherto unknown. Mr. Townend Glover in Washington, D. C., observed a *Toxophora*, the larva of which inhabits the well-known globular clay nest of the Wasp, *Eumenes fraterna,* "feeding either upon the caterpillar stored up in the nest, or upon the young larvæ themselves" (see Glover, Manuscript Notes from my Journal, etc., Diptera, p. 81, sub voce *Eumenes*). As far as I remember the specimen of *Toxophora,* which Mr. Glover kindly showed me many years ago, it was the species described below as *T. amphitea.*

The species described below may be tabulated as follows:

Second vein with a fork at the end, the posterior branch of which is connected by a cross-vein with the third vein :

The cross-vein between the discal and second posterior cells is S-shaped, and bears no stump of a vein..........1. *virgata* n. sp.

The cross-vein between the discal and second posterior cells is angular, and bears a long stump of a vein $\left.\right\}$ 2. *amphitea* Walk.
                                             3. n. sp. indescr. (Calif.).

Third vein with a fork at the end ; no cross-vein between the second and third veins.. ....4. *fulva* Gray.

T. amphitea and *virgata* have nearly the same venation; in both, it is the *second* vein, instead of the third, which is furcate; the posterior branch of the fork is connected by a perpendicular cross-vein with the third vein. This description applies, of course, to the venation as it appears to the eye; theoretically, it is the third vein, as usual, which is forked, the anterior fork being knee-shaped, and forming a square at the base, the anterior corner of which is connected by a *recurrent* cross-vein with the second vein, and thus produces the appearance of that vein being forked.

1. TOXOPHORA VIRGATA n. sp., ♂ ♀.—*Male.*—Head and antennæ black, second joint with a white reflexion on the inner side ; a tuft of white scales each side on the frontal triangle; occiput densely beset with pale yellow erect pile. The bluishblack ground-color of the thorax is more or less covered anteriorly with a fulvous tomentum and pale yellow pile; the pleuræ are covered with white, silvery scales ; thoracic bristles black. Abdomen black ; a stripe of ocher-yellow scales begins at the scutellum and reaches the end of the abdomen, being gradually attenuated ; the sides of the abdomen bear, on each segment, a large black spot, framed in by a ring of scales, which is yellowish on the dorsal side, more whitish toward the venter these rings being in close contact, their yellowish scales form, in well-preserved specimens, a longitudinal stripe parallel to the median dorsal stripe, and emitting, on the hind margins of the segments, branches of whitish scales, running toward the venter, which is another way of describing the same thing) ; venter densely clothed with white scales. Legs black; femora almost entirely, tibiæ partly clothed with white scales. Halteres with a yellow knob ; wings grayish or brownish, more or less tinged with yellowish in the costal, the first basal, and the inner end of the marginal cells ; a somewhat more saturate, almost brownish, spot on the præfurca; the cross-vein between the second posterior and discal cells is S-shaped, not angular, without stump of a vein (a vestige of one on one of the wings).

Female.—Front shining black, with some white scales on each side above the antennæ, the latter altogether black; in other respects, like the male.

Length about 7ᵐᵐ (measuring the chord of the curve formed by the body).

Hab.—Waco, Texas (Belfrage); Georgia (Morrison). Two males and two females.

2. Toxophora amphitea Walker, List., etc., ii, 298.

Head and antennæ black; second joint with a white reflexion on the inner side; a tuft of yellowish-white scales on the frontal triangle; occiput densely beset with pale yellow, erect pile. The black ground-color of the thorax is more or less covered anteriorly with a fulvous tomentum and pale yellow pile; the pleuræ are covered with white, silvery scales; thoracic bristles black. Abdomen black; a longitudinal stripe of scales along the dorsum gradually expands posteriorly; the scales upon it yellowish anteriorly become silvery-white posteriorly; on each side, the posterior margins have a short but broad cross-band of scales, yellow on the anterior, white on the posterior segments; these cross-bands are interrupted before reaching the dorsal stripe on segments 2–4; beyond the fourth segment, the cross-bands become more or less coalescent with that stripe; beyond the third segment, the cross-bands are coalescent with each other on the ventral side; venter with white scales. Legs black; femora with white and yellow, tibiæ with golden-yellowish scales. Wings as in *T. virgata;* but the cross-vein between the discal and second posterior cells is angular and bears a stump of a vein. Length about 5ᵐᵐ (measuring the chord of the curve formed by the body).

Hab.—Middle and Southern States. In preparing this description, I had two males from Kentucky and Georgia before me. The color of the covering of scales on the abdomen is very variable.

3. Toxophora spec., from California (H. Edwards).—Very like *T. fucata,* but larger and certainly distinct. I have only a single specimen, not well preserved enough to be described.

4. *Toxophora fulva* Gray, in Griffith's Animal Kingdom, xv, Insects, pt. ii, 779, tab. 126, fig. 5.

Ground-color opaque-black, but partly covered with fulvous scales and hairs. Face grayish-pollinose; front, in the female, covered with yellow scales; vertex with a few black bristles pointing forward; occiput with a dense fulvous fur. Thorax clothed with fulvous hairs on the front part of the dorsum; a fringe of shorter hair of the same color runs around the dorsum, the middle of which is usually denuded and black; scutellum also fringed with yellow hairs; thoracic bristles black. The posterior margins of the abdominal segments have borders of yellow scales, forming cross-bands, which coalesce on the sides of the abdomen; a longitudinal dorsal stripe of similar scales begins at the hind margin of the third segment, and runs to the end of the seventh. Venter for the most part clothed with yellow scales. Legs black, covered on one side with yellow scales (the posterior side on the four anterior legs and the anterior side on the hind legs). Wings brownish, yellowish along the anterior border; two submarginal cells; in other words, the geniculate anterior branch of the third vein is not connected by a cross-vein with the preceding vein ; the cross-vein at the distal end of the discal cell is bisinuated, but bears no stump of a vein. The antennæ of this species are comparatively more slender than those of *T. fucata* and *virgata ;* the first joint is clothed with yellowish-white scales.

Length across the curve of the body 8–9^{mm}. Straightened, the body would measure 10–12^{mm}.

Hab.—Georgia (H. K. Morrison). One male and two females.

In Griffith's Animal Kingdom, no *patria* is given; but, according to Walker (List, etc., ii, 298), the specimens came from Abbott's collecting in Georgia. The figure given in that work does not show the longitudinal yellow stripe on the abdomen ; nevertheless, the specific identity can hardly be called in doubt. As I observed before, no regular description is appended to the figure. I suspect that *T. leucopyga* Wied. and *T. fulva* are the same species.

EPIBATES nov. gen.

North America contains a number of species of a very elongated, almost *Thereva*-like shape, of a deep black color, and with rather long, distinctly 2-jointed, palpi. One of these species was referred by Macquart to the genus *Apatomyza* Wiedemann, the typical species of which is from the Cape. But already Walker (List, etc., iv, p. 1154), who had identified this species in Mr. Abbott's collection from Georgia, suggested that it belongs to a new genus. Although I have not the same species before me, I possess others which are evidently its congeners. The disagreement between them and Wiedemann's short description consists principally in the structure of the palpi, the last joint of which is not button-shaped, but lanceolate. The discrepancies in the venation, as figured by Wiedemann (tab. iv, f. 1), are only slight. But the general appearance of *Apatomyza punctipennis* on the figure is not that of the North American species above referred to. The abdomen in the latter is cylindrical, not tapering, as in the figure; the wings are longer, the head less close to the thorax ; the statement, " scutellum somewhat prolonged, with almost concave sides," finds no application to the American species; all the latter are distinguished by a deep black color, which is not the case with the species from the Cape.

For these reasons, we do not run any great risk in establishing a new genus, *Epibates*, for those American species. But these species, as far as known, have one character in common, which places their generic rights beyond any doubt, if it does not exist in the *Apatomyza* from the Cape. Four of the species before me of which I have male specimens have the thoracic dorsum beset with minute, rigid, sharp, conical points; this is apparently a sexual character, as it does not exist in my female specimens; unfortunately, none of my species is represented in both sexes.

The two genera recently described by Dr. Loew, and compared by him to *Apatomyza, Prorachthes* from Syria (Berl. Ent. Monatsschr., 1868, 380), and *Heterotropus* from Turkestan (Beschr. Eur. Dipt., iii. 180) are very different from *Epibates.* *Prorachthes* differs in the shape of the abdomen, the position of the head, the structure of face and front, of the first antennal joint, the venation; *Heterotropus* has short palpi and no spines on the legs. None of them has the peculiar muricate points on the thorax.

I have six species belonging to the same group, but unfortunately only one specimen of each; four of them are males, and two females; thus I am able to describe only one sex of each species; and this has to be borne in mind in reading my descriptions of the *generic* as well as specific characters. The presence of the sharp points on the thorax of the males of *E. luctifer, funestus, harrisi,* and *muricatus* proves that they are congeners; the position of *E. marginatus* and *magnus* in the same genus rests on characters taken from the other parts of the body. Macquart does not say anything of the presence of sharp points on the thorax of his *Apatomyza nigra* ♂; but they are easily overlooked in all the species except in the large *E. muricatus.*

Epibates, in Greek, means *a passenger.*

The characters of *Epibates* are as follows:—

Head on the same level with the thorax, and not much broader; occiput moderately convex, more so in the females (*E. marginatus* and *magnus*).

Eyes contiguous in the male for a short distance only, the apex of the vertical triangle being very much drawn out; ocellar tubercle distinct; the ocelli are placed on its sides, and for this reason, in the male, very difficult to perceive; in the two females, the eyes are separated by a broad interval; on each side of the ocellar tubercle, an ocellus is distinctly visible, but I do not perceive the third one (in *E. muricatus,* male, the eyes are separated by a very narrow interval).

Face and *lower part of front* subconically projecting in the profile; on the upper side of this projection, the antennæ are inserted; the interval between their base and the margin of the mouth (the epistoma or face) is narrow, sloping; the head descends but little under the eyes; oral opening oval, oblique, moderately large (Macquart's figure, Dipt. Exot., ii, 1, tab. 11, f. 1, *a,* showing the head in profile, is exaggerated, and the eyes are made to reach too low).

Proboscis longer by about one-half than the vertical diameter of the head; lips distinctly marked; palpi more than half as long as the proboscis; first joint ribbon-shaped, two or three times the length of the second, which is somewhat lanceolate.

Antennæ shorter than the head; first joint subcylindrical; second not much longer than it is broad, subcyathiform; the third about equal to the first in length, or a little shorter, flattened, somewhat lanceolate, attenuated at the end; terminal style none (in *E. marginatus* ♀, the antennæ are a little longer than the head, and the first joint is distinctly longer than the third).

Thorax but little convex; its dorsum, in *E. funestus, luctifer, harrisii,* and *muricatus,* of which I have only male specimens, is beset with minute, rigid, sharp, conical points, arranged in irregular rows. As *E. marginatus* and *magnus,* of which I have only females, do not have these points, it seems very probable that this is a sexual character.

Scutellum comparatively large, almost semicircular, convex, cushion-shaped.

Abdomen cylindrical, long and slender, by one-half longer than head and thorax taken together; in the male, eight segments, the genitals forming the ninth; in the females of *E. marginatus* and *magnus,* I count only seven segments besides that bearing the genitals.

Legs long and slender, beset with sparse spinules along the tibiæ; hind legs by far the longest; *pulvilli* distinct, rather broad; *ungues* curved, broad at base. In my female specimens, I perceive a few stiff spine-like bristles on the under side of the hind femora, two in *E. marginatus;* four or five in *E. magnus.* I do not see anything like it in the males.

Wings but little shorter than the body, rather narrow, attenuated at the base; alula small, very narrow. In *E. muricatus,* the wings are broader.

Venation.—Two submarginal and four posterior cells; first posterior broadly open; upper branch of third vein gently S-shaped, inserted about the middle of the section of that vein beyond the small cross-vein. The latter corresponds to the middle of the discal cell. Præfurca less than half as long as the distance between the bifurcation and the small cross-vein; second vein gently arcuated on its latter half, reaching the margin without forming any sinus; thus the marginal cell is not expanded at its end. The proximal end of the third posterior cell is opposite the small cross-vein; anal cell open (Macquart's figure, Dipt. Exot., ii, 1, tab. 11, f. 1, gives a tolerably correct idea of the venation, except that the anal cell is represented as being closed; the upper branch of the third vein in *E. magnus* and *muricatus* is nearly as bisinuate as represented, but it is less so in the other species). The costal margin in the male sex is beset with minute blunt points, as in *Ploas;* they are almost obsolete in some species (*E. funestus*); very distinct in others (*E. muricatus*).

In the following table I include Macquart's *E. niger* from the data in his description :

Wings infuscated, but anal angle (including at least the second basal, anal, and axillary cells) hyaline :

Small species (7–8^{mm} long):

      1. *funestus* (♂).—White Mountains, N. H.

Large species (12–14^{mm} long):

    Prevailing pubescence black :

      5. *magnus* (♀).—Vancouver Island.

Prevailing pubescence gray :

6. *harrisi* (♂).—Atlantic States (?)

Wings, including the anal angle, infuscated :

Well-marked brown clouds on the cross-veins, and at the bifurcations :

7. *niger* ( ).—Georgia.

Brown clouds, etc., indistinct or none :

Abdominal segments with a fringe of whitish hairs posteriorly :

4. *marginatus* (♀).—California.

No fringes of whitish hairs on the abdominal segments, which
are deep velvet-black :

Small species (8^{mm}); stem of halteres pale; knob brown :

2. *luctifer* (♂).—Vancouver.

Large species (15^{mm}); halteres altogether blackish :

3. *muricatus* (♂).—Sierra Nevada.

1. EPIBATES FUNESTUS n. sp., ♂.—Deep velvet-black ; wings dark
brown along the anterior margin, posteriorly hyaline on the proximal,
brownish-hyaline on the distal half. Length 7.5^{mm}.

Head, antennæ, proboscis, and palpi black ; the frontal triangle and
the orbits of the eyes, in a certain light, have a white reflection ; under
side of the head with long, white pile ; oral edge, vertex, and occiput
with long black hairs. Thorax deep velvet-black, opaque, beset on the
dorsum with minute, sharp, rigid points, and sparse, long, black pile ;
pleuræ and coxæ clothed with grayish pollen, and sparsely beset with
white and blackish hairs. Abdomen deep black, opaque, with some
scattered pile on the lateral and under side, which is white at the base
and black beyond it. Halteres whitish-yellow, with a brown knob.
Legs black. Wings dark brown along the anterior margin as far as the
apex, including the first basal, marginal, and two submarginal cells ;
from the latter posteriorly, the brown becomes gradually evanescent,
until it almost disappears in the last posterior cell ; second basal, anal,
and axillary cells hyaline, or almost so ; the surroundings of the anterior cross-vein are darker brown, those of the stigma still more so ; on
the other cross-veins and on the bifurcation of the third vein the clouds
are almost obsolete. The denticulation of the costa is almost obsolete.

Hab.—White Mountains (H. K. Morrison). A single male in Mr. E.
Burgess's collection.

2. EPIBATES LUCTIFER n. sp., ♂.—Deep velvet-black ; wings uniformly brown. Length 8^{mm}.

Resembles the preceding in its coloring, but is easily distinguished
by the uniformly brown color of the wings, which is only slightly darker
on the distal half anteriorly and around the small cross-vein ; denticulations on the costa minute, but distinct. Remains of some short, reddish-golden pile are perceptible on the sides of the thoracic dorsum,
especially above the root of the wings. The thorax is beset with the
same minute, sharp, rigid points ; the under side of the head with long,

white hairs, which are also found on the front coxæ and the pleuræ; halteres with a brown knob. Legs black, in a certain light with a purplish reflection. *Au ʃʃ· · ·ʃ · ʃ· µʃ·ʃ·ʃ·*   *?*

Hab.—Vancouver Island (G. R. Crotch). A single male.

3. EPIBATES MURICATUS n. sp., ♂.—Deep velvet-black; wings infuscated; halteres, including the stem, black; denticulations of the costa large and distinct. Length 15^{mm}.

Whole body uniformly deep black, very opaque on the thorax; all the pile black, including that on the under side of the head. Halteres altogether black. Front tarsi somewhat brownish. The sharp, rigid points with which the thoracic dorsum is beset are very distinct, and of different sizes; some are quite large; short, black pile among them. The wings are comparatively broader here than in the two preceding species; they are tinged with brown, the centers of the cells being somewhat paler; the latter portion of the costal cell dark brown; small cross-veins somewhat clouded; anterior branch of the third vein strongly bisinuate, almost S-shaped; discal cell somewhat broader and shorter than in *E. funestus* and *luctifer;* posterior cells 2, 3, 4, and especially the last, much shorter, the posterior cross-vein being a little farther from the proximal end of the discal cell.

Hab.—Sierra Nevada, California (H. Edwards). A single male.

4. EPIBATES MARGINATUS n. sp., ♀.—Black, beset with whitish pile, which forms fringes on the hind margins of the abdominal segments; wings infuscated. Length 8^{mm}.

Antennæ comparatively longer than in *E. funestus*, owing to the length of the first joint, which is distinctly longer than the third; they are black, beset with black pile. Head black, slightly whitish-pollinose along the inner orbits, beset with black pile on face and front, and with white pile on the under side and occiput; face rather projecting. Thorax black, subopaque, beset with long whitish pile, more dense and of a purer white on the pleuræ and coxæ. Abdomen deep velvet-black; hind margins of the segments with a fringe of short whitish pile; the last segment smooth, shining. Halteres with a brownish-white stem and brown knob. Legs black. Wings tinged with blackish-brown, which is more saturate along the veins, so that the inner portion of the cells is paler; the darker color of the base and of the anterior margin is very gradually evanescent posteriorly. The hind femora, on the under side, besides the usual pile, have two minute, stiff, spine-like bristles.

Hab.—San Francisco, Cal. (H. Edwards).

5. EPIBATES MAGNUS n. sp., ♀.—Altogether black; thorax deep velvet-black, with black pile; occiput beset with yellowish-white pile; first abdominal segment with a fringe of white pile; wings tinged with brown; second basal, anal, and axillary cells hyaline. Length 12–13^{mm}.

Head beset with black pile, except on the occiput, where it is yellowish-white; front very little shining, with a faint trace of grayish pollen.

Thorax deep velvet-black, with black, moderately long, erect pile; pleuræ somewhat shining, and, in a very oblique light, slightly grayish pollinose. Abdomen moderately shining, beset with sparse black pile, which is more dense on the under side and round the tip; first segment posteriorly with a fringe of white pile; some few white hairs on the side of the second segment on its hind margin. Halteres dark brown, the stem yellowish-brown. Wings infuscated, but less so than in the other species of the genus, darker along the anterior margin, the region of the stigma dark brown; brown clouds on the cross-vein at the base of the fourth posterior cell and on the bifurcation of the third vein, but both very little conspicuous; second basal, anal, and axillary cells hyaline; the fifth vein is margined with brown, the sixth is not. The hind femora on the under side, besides the usual pile, have four or five stiff, spine-like bristles.

Hab.—Vancouver Island (G. R. Crotch).

6. EPIBATES HARRISI n. sp., ♂.—Black; thorax and abdomen clothed with long, grayish-white pile; wings hyaline, anterior margin brown (the root, costal, first basal, and marginal cells). Length 14^{mm}.

Epistoma black, shining; occiput thickly clothed with grayish-white pile. Thorax deep black, opaque, clothed with whitish-gray pile, sparsely on the dorsum, more densely on the pleuræ; the rigid, sharp points on the dorsum are distinct. Abdomen cylindrical, black, with grayish-white recumbent pile, rather uniformly spread over all the segments (on the hind margins of each of the segments 3–7 in the middle, there is a small spot, denuded of gray pile, and therefore appearing darker black; these spots are too regular to be an accidental denudation of the specimen, which is nevertheless possible); on the sides of the first four segments long, white, erect pile; on the sides of the following segments, beginning with the fifth, a fringe of long, erect, black pile. Femora black, with some white pile; tibiæ and tarsi dark brown. Halteres yellowish-white, with a brown knob. Root of the wings, costal, first basal, marginal, and the proximal end of the first submarginal cells brown, somewhat darker around the stigma and on the small cross-vein; the remainder of the wing hyaline; veins brown; anterior margin with distinct denticulations.

A single male, in T. W. Harris's collection in the Boston Museum of Natural History; a label with H. Gray upon it. It is very probably from the Northern United States, as nearly all the specimens of the collection.

7. EPIBATES NIGER Macquart, Hist. Nat. Dipt., i, 390, 2 (*Apatomyza*); Dipt. Exot., ii, 1, 111, tab. xi, f. 1 (*id.*).

"Length 4½ lines. Black, with gray pile. Palpi reaching the tip of the antennæ; first joint elongated, cylindrical, hairy; the second a little less long, glabrous, attenuated at the base and tip. Wings spotted as in the preceding species (Wiedemann's *Apatomyza punctipennis* from the Cape); anterior margin denticulate on its posterior half in the male."

Hab.—Georgia.

Observation.—This paper was already in press when I received a specimen of *Epibates* (*Apatomyza*) *niger* Macquart, collected by Mr. H. K. Morrison in Georgia. It is a male, but a not very well preserved specimen. The little spines on the thorax are almost obsolete in this species; with a strong magnifying glass, some traces of them are visible. The eyes are not contiguous on the front, but, like those of *E. muricatus* ♂, separated by a narrow, linear interval. The third antennal joint is much broader than in the other species; it expands immediately beyond the base, contracts again about the middle, and ends in an elongated point; the curves it forms, above and below, are not quite symmetrical, the one below being flatter; altogether, it has the shape of an elongated and somewhat irregular ace of spades. The principal figure of Macquart's gives a somewhat more correct representation of it than the figure of the head in profile. The anal cell is open.

Family THEREVIDÆ.

PSILOCEPHALA COSTALIS Loew, Centur., viii, 16.—California.
THEREVA COMATA Loew, Centur., viii, 9.—California.
THEREVA FUCATA Loew, Centur., x, 37.—California.
THEREVA MELANONEURA Loew, Centur., x, 36.—California.
THEREVA HIRTICEPS Loew, Berl. Entom. Zeitschr., 1874, 382.—San Francisco.
XESTOMYZA PLANICEPS Loew, Centur., x, 38.—California.

California seems to be quite rich in *Therevidæ*, as I have collected four species of *Thereva* (two in Marin County, one in Southern California, and one in Yosemite Valley), none of which I am able to identify with the above quoted descriptions. *Xestomyza planiceps* I received from Mr. Henry Edwards. The following snow-white *Thereva* was very common in Yosemite Valley about the beginning of June :—

THEREVA VIALIS n. sp., ♂.—Grayish; clothed with snow-white pile, especially on the abdomen; antennæ black; femora black, with gray pollen; tibiæ brownish-yellow, black at tip; tarsi dark brown, brownish-yellow at base. Length 8–9^{mm}.

Male.—Head white, with white pile; some black bristles on vertex, and alongside of them on the occiput; some others on the face on each side of the antennæ (in some specimens only a few, which, for this reason, are discernible with difficulty); antennæ black, first joint not longer than the two following together, slender, whitish-pollinose, with white pile and near the tip with some black bristles. Thorax gray, being clothed with a dense pollen; two distinct longitudinal stripes white; white pile, especially on the pleuræ; the ordinary bristles black. Abdomen densely clothed with silvery-white pollen and long white pile; a few black bristles on the under side of the hypopygium. Femora dark, clothed with gray pollen and white pile; tibiæ brownish-yellow, the tip black; tarsi dark brown or black, base of the first joint brownish-yellow; on the middle tarsi, this color occupies nearly three-quarters of the joint. Knob of halteres white, black at base. Wings hyaline; veins brown, except those at base and near the costa, which are pale yellow; fourth posterior cell closed.

Hab.—Yosemite Valley, California, June 9–11, common. Seven males.

Is very like *Thereva candidata* Loew from the Atlantic States, but differs in having a few black bristles on the face each side of the antennæ; the femora dark to the very tip; the tarsi brown, except at base; the wing-veins darker; the third antennal joint likewise darker.

Family SCENOPINIDÆ.

As I have no Californian species of this family, I will describe the following remarkable species from Missouri:—

SCENOPINUS BULBOSUS n. sp., ♂ ♀. *First posterior cell closed*, petiolate; head, thorax, and the sides of the abdomen sparsely covered with coarse, pollen-like grains. Length 5–5½^{mm}.

Antennæ black, hardly reddish at the suture between the second and third joints. Head and thorax blackish-bronze color; the front ♂ is an acute triangle, meeting the triangle of the vertex; the line of contact of the eyes is thus a very small one; in the ♀, the eyes are not contiguous, but separated by the moderately broad front; both front and vertex, in both sexes, are sparsely covered with yellowish-white, coarse, pollen-like grains. Thorax stouter and more gibbous than in *S. fenestralis*, covered above and on the sides with the same pollen-like grains, which are not dense enough, however, to conceal the ground-color. Abdomen black above; its sides and the venter covered with the same grain-like pollen. Halteres brown; legs black; roots of the tarsal joints more or less yellowish. Wings subhyaline (♂), slightly brownish anteriorly (♀); costal cell brownish; first posterior cell closed, the fourth vein being incurved toward the third, and ending in it at a considerable distance from the margin of the wing; the petiole thus formed is about equal in length to the posterior transverse vein in the ♂; a little shorter in the ♀. The second submarginal cell is nearly as long as the first posterior (therefore much longer than in *fenestralis*); the distance between the two cross-veins is a little shorter than the great cross-vein.

Hab.—Missouri, in July (C. V. Riley).

The grains of pollen which distinguish this species appear, under a magnifying power of 100–150, like elongated bulbs inserted on short stalks.

Obs.—This species shares the closed first posterior cell with the new genus *Atrichia*, formed by Loew for a Mexican Scenopinid (Centur., vii, 76); the latter, however, is described as elongated, slender, with slender feet, characters which by no means belong to *S. bulbosus*. The name *Atrichia*, revived by Dr. Loew from Schrank's Fauna Boica, 1803, where it was used for *Scenopinus*, cannot be maintained, as, in the mean time, the same name has been used by Mr. Gould, in 1844, for a genus of

Birds. As *Atrichia* Loew is not the same thing as *Atrichia* Schrank, and cannot, for this reason, date its claim earlier than 1866, *Atrichia* Gould has the priority. I propose to call the genus *Pseudatrichia*.

Family CYRTIDÆ.

The species described here are:—One *Opsebius* from California, and a second one from Vancouver Island; a *Pterodontia* from Oregon; two new *Eulonchi*, which raises to four the number of species of this peculiarly Californian genus. An *Oncodes*, which I also possess, has been already described by Mr. Loew.

The descriptions of a large new *Ocnæa* from Texas and of an *Oncodes* from New England are also added.

EULONCHUS.

Established by Gerstaecker, in the Stett. Ent. Zeitschr., 1856, for *E. smaragdinus*, this genus has been gradually increasing since, and counts now four species. None of them, as far I know, have been found outside of California. Within that State, they occur almost at sea-level, on the sands of Lone Mountain, San Francisco, as well as at an altitude of 8,000 feet in the Sierra Nevada.

Legs altogether yellow:
    Proboscis (in   and ♀ longer than the abdomen; body of the female
        bright metallic green....................... *smaragdinus* Gerst.
    Proboscis shorter than (♂) or as long as (♀) the abdomen; body
        metallic blue or purplish in both sexes........ *sapphirinus* n. sp.
Legs, or at least femora, black:
    Tip of femora and the greater part of the tibiæ whitish-yellow;
        tegulæ uniformly white........................... *tristis* Loew.
    Legs altogether black, only the knees paler; tegulæ margined with
        black .... *marginatus* n. sp.

1. EULONCHUS SMARAGDINUS Gerstaecker, Stett. Ent. Zeit., 1856, p. 360.—Not uncommon on the sands about Lone Mountain, San Francisco, according to the statement of Mr. H. Edwards. Three green specimens which I have are females. Two male specimens which I received from Mr. Edwards are smaller (one of them only 9–10ᵐᵐ long), the proboscis shorter, although still exceeding the abdomen in length; the coloring is bluish on the thorax, purplish-blue on the abdomen. Are they the males of this species? If they are, Dr. Gerstaecker was mistaken in describing his green individuals with a long proboscis as males.

2. EULONCHUS SAPPHIRINUS n. sp.—Antennæ black, sometimes brownish or reddish toward the tip; epistoma black or bluish-black; ocellar triangle dark blue or purple; sheath of the proboscis black; body metallic blue or purple, sometimes with greenish reflections, clothed with dense, erect, grayish-yellow pile on the thorax; abdomen with similar but much less dense pile, and with an appressed yellowish-white pu-

bescence, visible in a certain light only; feet straw-yellow; tarsi brownish toward the tip; wings grayish-subhyaline; costal cells brownish-yellow; costal and first longitudinal veins black on their proximal half, brownish-yellow toward the end; tegulæ whitish, their margins yellowish; knob of halteres yellow. The proboscis of the male does not reach the end of the abdomen; that of the female does not reach beyond it. Length 9–11^{mm}.

Hab.—Webber Lake, Sierra County, California, July 23–26, not rare, flying in circles around flowers. Three males and two females. A male and a female from Calaveras, Sierra Nevada, California, June (G. R. Crotch), have the proboscis a little longer than the abdomen.

This species is easily distinguished from *E. smaragdinus* ♀ by its smaller size, blue color, shorter proboscis, less yellowish wings; the two latter characters also distinguish the males, which are somewhat alike in coloring.

All my specimens, as far as I remember, were more uniformly blue when I took them, and seem to have assumed the purple and even greenish tinges, which they have now, in the process of drying.

3. EULONCHUS TRISTIS Loew, Centur., x, 19.—I found a male and a female in the Coast Range, in the woods of *Sequoia sempervirens*, above Santa Cruz, Cal., on a flower, May 21, 1876.

4. EULONCHUS MARGINATUS n. sp.—Metallic green, with bluish reflections on the scutellum, the anterior margins of the segments, etc.; venter metallic blue. Antennæ black. Thorax clothed with dense pale yellowish-white erect pile; abdomen with a short appressed pubescence, which forms whitish cross-bands along the hind margins of the segments. Legs black, and only the knees yellowish-white. Tegulæ with very distinct black margins. Wings subhyaline; all the veins dark brown, except the distal end of the costa and of the first posterior vein, which are reddish-yellow. Proboscis a little longer than the abdomen. Length 9^{mm}.

Hab.—Napa County, California (H. Edwards). A single specimen, apparently a male. The petiole of second submarginal cell is subobsolete; as I have only one specimen, I cannot say whether this is a permanent character of the species.

PTERODONTIA MISELLA n. sp.—Black; clothed with black pile; scutellum black, obscurely reddish on its latter half; second abdominal segment (that is, the first visible segment; the true first segment is concealed under the scutellum) black, with an obscurely marked reddish spot on each side a little back of the scutellum; segments 3–6 rufous, the third and fourth with square black spots in the middle, that on the fourth being narrower; they are confluent with each other and with the black of the second segment. Venter rufous; hind margins of segments 2–5 black. Tegulæ brownish, with broad dark brown margins. Legs brownish-yellow, the four posterior femora black; ungues reddish at base, black at tip. Wings subhyaline; veins yellow; venation similar to that of the other species; the usual tooth on the edge of the costa, near the end of the first posterior vein, is very little projecting. Length 5^{mm}.

Hab.—Oregon (H. Edwards). A single specimen. This species is very like *P. flavipes* from the Atlantic States, but is smaller and differs in the coloring of the abdomen.

LASIA KLETTI Osten Sacken, in Lieutenant Wheeler's Report Explorations and Surveys, etc., vol. v, Zoölogy, 804.—Arizona.

OCNÆA HELLUO n. sp.—Two submarginal cells; five posterior cells, the first divided in two by a cross-vein, and the second half of it closed and petiolate, the fourth posterior cell likewise closed and short-petiolate; all the longitudinal veins reach the margin; body black, beset with short yellowish pile; hind margins of the abdominal segments with broad yellow borders, expanding along the lateral margins; legs yellow, including the coxæ. Length 13–14^{mm}.

The venation is like that of *O. calida* (Wiedemann, Auss. Zw., ii, tab. vii, f. 2*b*), with the following modifications:—1. The third vein emits a branch some distance beyond the cross-vein dividing the first posterior cell; thus a second submarginal cell is formed; 2. The cross-vein in the first posterior cell is just opposite the cross-vein at the base of the second posterior cell, and not far beyond it, as in Wiedemann's figure; 3. The vein between the second and third posterior cells reaches the margin; 4. The fourth posterior cell, which is closed, is much longer, forming an irregular parallelogram, with a cross-vein at its base.

Antennæ dark-brown, basal joints reddish, the elongated third joint also somewhat reddish on the inner side. Thorax black, shining, clothed with dense and soft yellowish-gray pile, almost rendering it opaque; humeral callosities whitish-yellow; antealar callosities brownish. Abdomen black, densely clothed with short, erect, yellow pile; all the segments with broad clay-yellow hind borders, expanding laterally so as to occupy the whole lateral margin; ventral segments black, with broad clay-yellow hind borders. Legs including coxæ yellow, the extreme end of the last tarsal joint and the ungues black. Wings very slightly tinged with brownish; costal cell a little more saturate.

Hab.—Dallas, Texas (Boll). One specimen.

Observation.—This fine species is not unlike Erichson's figure of *O. longicornis* (Entomographien, tab. i, f. 8), but the venation is different, the black on the abdomen occupies more space, the hind tibiæ are brown, the abdomen much stouter; the size is larger by one-half than the hair-line of the figure.

OPSEBIUS DILIGENS n. sp.—Of a slightly metallescent brownish-black color, clothed with brownish-yellow pile; legs brownish-yellow; wings tinged with brownish, the tip hyaline; first posterior cell divided in two by a cross-vein; the bases of the third and fourth cells nearly on the same line; anal cell closed and petiolate. Length about 5^{mm}.

The venation is like that of the European *O. inflatus* Loew (Wiener

Entom. Monatsschr., 1857, p. 33, tab. i, f. 1), with the following differences:—1. The first posterior cell is divided in two (nearly equal) parts by a cross-vein placed between the end of the discal and the proximal end of the second submarginal cell (the same character distinguishes the two North American *Opsebius* described by Mr. Loew in the Centuries); 2. The third and fourth posterior cells have their proximal ends nearly on the same line; in other words, the insertion of the intercalary vein is coincident with the cross-vein at the base of the fourth posterior cell; 3. The fifth vein runs straight to the margin, and the sixth is incurved toward it a short distance from the margin. The costa is distinctly thickened between the ends of the first and the third veins, and a little beyond the latter. The wing is distinctly tinged with brownish, except at the base and the tip, which are subhyaline.

Body of a uniform brownish-black, slightly metallescent on the thorax. Thorax densely clothed with brownish-yellow erect pile, not dense enough, however, to conceal the shining surface under it. On the abdomen, the same pile is more dense on the second segment; the pile on the two intermediate segments is more blackish, except along the posterior margins, where it is yellowish; the fifth has a shorter and more appressed whitish-yellow pubescence, interspersed with longer pile; the last segment is black, shining, transversely rugose. Legs brownish-yellow; femora slightly tinged with brownish; coxæ, except the extreme tip, brown. Halteres with a yellowish-white knob; tegulæ semitransparent, colorless. Eyes pubescent; antennæ (broken).

Hab.—Vancouver Island (G. R. Crotch). Two specimens.

OPSEBIUS PAUCUS n. sp.—Very like *O. diligens*, but smaller, 4–5ᵐᵐ long; sixth vein interrupted before the nearest cross-vein, and thus the anal cell open; the branches of the fourth vein do not quite reach the margin. Antennæ yellowish-brown at base; distal portion of the last joint and arista nearly black; pubescence of the eyes long and dense. Thorax with very dense, soft, erect, grayish-yellow pile; the greenish-black, shining ground-color but little visible under it. Abdomen brownish-black, moderately shining, densely clothed with brownish-yellow erect pile; the penultimate segment and the hind margin of the preceding one are clothed with recumbent yellowish-white pile. Wings slightly tinged with brownish, much less than in *O. diligens*, but more uniformly, as the paler color of the tip is not apparent. The rest as in *O. diligens*.

Hab.—California (G. R. Crotch). One specimen.

ONCODES MELAMPUS Loew, Centur., x, 17.—California. I have a specimen (brought by G. R. Crotch) which I doubtfully refer here. The tibiæ are brown, not black; the borders of the tegulæ very pale brownish; the wing-veins are very pale, except those near the costa, which are brownish.

ONCODES INCULTUS n. sp.—Brownish-black; humeral callosities brownish-yellow; antescutellar callosities yellowish-brown; posterior margins of abdominal segments white; legs dark brown; knees brownish-yellow; wings strongly tinged with brown. Length 8^{mm}.

The brownish-black thorax and scutellum are clothed with a dense, short, yellowish pubescence; abdomen dark brown, segments 2 and 3 with narrower, 4 and 5 with broader, white posterior margins; venter, except the base, white; each segment with a black cross-band on the anterior margin. Tegulæ brownish, with narrow dark brown edges. Halteres with a brown knob. Wings comparatively long, strongly and rather uniformly tinged with brown. This color is darker in the costal cells, especially in the interval between the auxiliary and the first veins; costa distinctly incrassate in the region of the stigma; veins brown.

Hab.—White Mountains, New Hampshire. Two specimens.

Easily distinguished from the other described species of the genus by its strongly infuscated wings and its large size. The abdomens of my specimens being somewhat shrunken, the measurement I give is only an approximation.

Family MIDAIDÆ.

We have two Californian *Leptomidas*, the eight other species of the genus belonging to the Mediterranean fauna (Portugal, Algiers, Egypt) or South Africa.

The new genus *Rhaphiomidas* from California is closely related to *Mitrodetus* from Chili.

The anomalous genus *Apiocera*, intermediate between the *Midaidæ* and *Asilidæ*, has been found in Australia, Vandiemen's Land, and Chili. I describe a species from Yosemite Valley, California.

LEPTOMIDAS PANTHERINUS Gerstaecker, Stett. Ent. Zeit., 1868, 85.

(Translation.)—"Wings slightly infuscated, with testaceous veins; body, antennæ, and legs luteous; head and basal cross-bands on the abdominal segments black. Length 8⅔ lines" [about 19^{mm}].

"Antennæ considerably longer than the thorax; the stout basal joint only twice as long as the second, both beset with black bristles; the third joint is twice as long as the two first taken together, with an incrassate, distinctly separated tip; the terminal club is equal to the whole antenna in length, and is divided by a coarctation on the first third of its length in a narrow basal and an elongate-oval apical portion. The color of the antennæ is reddish-yellow; the tip of the third joint and the base of the terminal club are infuscated. Head black, only the small tumid clypeus and the lower oral edge reddish-yellow; pubescence altogether golden-yellow, somewhat longer on the clypeus. The broad labella of the short proboscis reddish-brown. Thorax yolk-yellow, with indistinct darker stripes and short yellowish pile on the dorsum; pleuræ shining light brown, variegated with black. Legs uniformly yellow, with light brown coxæ; ungues black at tip; hind femora not incrassate, before the tip on the inner side with a chestnut-brown longitudinal streak, on the lower side sparsely beset with rather thin spines; hind tibiæ straight, long, and slender, sparsely beset with spines on the inner side as far as the knee; at the tip, a circle of longer spines.   Wings uniformly tinged with a diluted brownish; veins pale yellow; halteres yellow.   Abdomen of the same ground-color as the thorax, but with black cross-bands; the first segment quite black, except the hind margin, the second black on the anterior half; segments three to six have basal black cross-bands, triangularly expanded in the middle; they become narrower on each consecutive segment, so that the cross-band of the sixth segment is only a narrow anterior border; the hind margin of the segments is paler, more straw-yellow; on the second segment, it has on each side a brown, transverse callosity; the seventh and eighth segments are a little darker than the preceding ones, and are densely beset with black bristles; the spines of the last segment are ferruginous, obtuse.

"A single female from California, in the Berlin Museum."

I took two females on the sands of Lone Mountain, San Francisco, June 29, 1876.   The coloring of the antennæ is variable; in one of my specimens, the whole club is black; in the other, it answers Dr. Gerstaecker's description; in both of my specimens, the knob of the halteres is brown, and not yellow.

LEPTOMIDAS TENUIPES Loew, Centur., x, 20.—California.

MIDAS VENTRALIS Gerstaecker, l. c., 102 (syn. *M. rufiventris* Loew, Centur., vii,₋22).—California.

I do not know these species.

RHAPHIOMIDAS nov. gen.

Closely allied to *Mitrodetus* Gerstaecker (Stett. Ent. Zeit., 1868, 76), as there are three cells intervening between the forked cell and the margin of the wing, and as the structure of the proboscis is the same, long and linear, directed forward, with very narrow lips at the end; differing, however, from that genus in the structure of the antennæ; in some minor characters of the venation, among them the structure of the second submarginal or forked cell, which is petiolate at the proximal end only, and not at both ends; and in the presence of two distinct ocelli.

Vertex somewhat excavated on each side of the tubercle; the latter broad and flat, bearing two large *ocelli* on its sides (a character, as far as I am aware, unique among the true *Midaidæ*).

Antennæ a little longer than the vertical diameter of the head from the top of the eyes to the lower oral edge, inserted rather low, a short distance above the mouth; the first two joints form an almost cylindrical body, somewhat constricted about two-thirds of its length where the second joint begins; third joint about once and three-quarters of the two first taken together, in the shape of a rabbit's ear, with a ring-like expansion at the basis.

Face very short, in the profile nearly straight, moderately advancing in front of the eyes; oral edge cut obliquely; cheeks moderately broad.

Proboscis, if bent backward, would reach the scutellum, linear, straight, pointing forward; the two narrow lips at the end a little curled up. In my specimen, the head is somewhat distorted from its natural position; the proboscis is longitudinally cleft in two parts, both long and linear, at an angle to each other. Macquart (Dipt. Exot., 4e suppl., tab. iv, f. 1) represents the proboscis of *Mitrodetus* [*Cephalocera*] *dentitarsis* in a somewhat similar manner. Nevertheless, owing to the imperfect condition of my specimen, I am not prepared to affirm that this is a permanent character.

Venation of the wings:—Three cells intervene between the second submarginal cell and the margin of the wing. That cell is petiolate at the proximal end, coarctate at its distal end, which coincides with the tip of the first vein. The first posterior cell comes in contact with the first basal cell (I mean to say, is not petiolate). The small cross-vein near the posterior margin is absent, although a rudiment of it, in the shape of a minute stump of a vein, is perceptible in the usual place. (The venation is not unlike that in Gerstaecker, l. c., tab. i, f. 1; but the second submarginal cell is more ventricose, the contact between the first posterior and first basal is broader, so that the angle of the latter is not projecting; the discal-cell is shorter and broader, the second basal longer, the small cross-vein on the posterior margin absent, etc.)

In front of the halteres, there is a singular conical body, a little shorter than the halteres, the homology of which I do not attempt to explain.

Abdomen of the female with a circle of spines at the end.

I possess only a single, very much injured, female specimen; and if I venture, nevertheless, to describe it, it is on account of its very marked generic characters and its evident relationship to the Chilian genus *Mitrodetus*.

RHAPHIOMIDAS EPISCOPUS n. sp., ♀.—My only specimen having been very much injured by moisture, I can say very little about its natural color; at present, it is uniformly black, opaque (originally, it may have been gray); the three last abdominal segments shining; remains of long, brownish-yellow pile are visible on all parts of the body; short, black, appressed pile on the three last abdominal segments; knob of the halteres yellowish-brown. Antennæ dark brown, the third joint reddish-brown, especially at the base. Front coxæ black, beset with long yellow pile; femora dark brown; tibiæ reddish-brown, with dense, recumbent yellowish pile, and some scattered, long, black bristles; tarsi brownish-red; hind femora black (the middle legs and the hind tibiæ are broken off). Wings subhyaline, with a slight brownish tinge; the costal cell before the humeral cross-vein saturate yellowish-brown; the extreme proximal end of the marginal cell and the distal end of the costal cell have a similar brownish tinge. Length about 25^{mm}.

Hab.—California. One female.

The specimen has been for many years in my collection, labeled "California". I do not remember from whom I received it, but it may have come with a small lot of insects from Lower California.

APIOCERA.

A genus of doubtful systematic position; refused admittance to the *Midaidæ* by Dr. Gerstaecker, the last monographer of the family; excluded by Dr. Loew from the *Asilidæ*; not less remarkable for its geographical distribution.

Westwood (Lond. and Edinb. Phil. Mag., 1835; Arcana Entomologica, vol. i) introduced it for three species from Australia, and referred it, with a doubt, to the *Midaidæ*.

Macquart (Dipt. Exot., 2e suppl., 49, tab. ii, f. 1, 1847) introduced the same form from Van Diemen's Land, under the name of *Pomacera*, establishing a separate family, *Pomaceridæ*, for it.

Philippi (Verh. zool.-bot. Ges., 1865, 702, tab. 25, f. 26) established the genus *Anypenus* for the same form, discovered in Chili; he places it among the *Asilidæ*, and describes two species.

I possess a species from California which is undoubtedly an *Apiocera*. It has the same large, broad, spoon-shaped palpi; a short, strongly retreating face; a proboscis with very large lips; antennæ with a short, somewhat pear-shaped terminal joint, bearing a small style; the venation is exactly like that represented by Philippi in the above-quoted figure; the character of the coloring is the same as that of all the previously described species.

APIOCERA HARUSPEX n. sp., ♂.—Blackish-gray; abdomen black, with white cross-bands; three segments preceding the black hypopygium are white. Length 20^{mm}.

Face and palpi white, beset with white pile; antennæ black, basal joint beset with long white pile; front white; ocellar tubercle blackish. Thorax grayish-black above; humeral callosities white; a whitish longitudinal line and two lateral lines on the dorsum; the latter are expanded anteriorly into broad white triangles; two other white lines, curved anteriorly, between these lateral lines and the pleuræ; pleuræ grayish-white. Abdomen, first segment whitish on the sides, brownish in the middle, and with a fringe of black pile posteriorly; anterior margin of the second segment with a white cross-band, emarginate in the middle, expanding laterally; posterior margin with two large contiguous white triangles, prolonged laterally so as to coalesce on the lateral margin with the anterior cross-band; the intermediate region of the segment is deep black, opaque; third segment black, opaque, with a white cross-band on the anterior margin, emarginate in the middle; fourth segment black, with a vestige of a narrow white margin anteriorly, concealed under the preceding segment; the three following segments white; hypopygium black, large, oblong, resembling that of an *Erax*. Venter white. Legs grayish, with black spines; tibiæ and tarsi reddish-brown.

Wings hyaline; veins dark brown; venation exactly like that in Mr. Philippi's figure of *Anypenus*.

Hab.—Yosemite Valley, California.   One male specimen.

Family ASILIDÆ.

In working up my western materials in the family *Asilidæ*, I have paid especial attention to the section of the *Dasypogonina*, as the most numerous and the most rich in peculiar generic forms. I also described a few of the more striking forms of *Laphrina*. The section *Asilina* I have altogether left out for the present, for the reason that the *Asilina* from the Atlantic States are still in a state of confusion, and it will be better to work up all the species of this difficult group together.

My Californian collection in the section *Asilina* is remarkably small, which may in part be accounted for by the fact that these flies were not in season yet when I left San Francisco for the Sierra Nevada in the middle of July.   Up to that date, I had only found a single species of *Machimus* (the Geysers, Sonoma County, May 5–7; also in Mariposa County, and in Yosemite Valley, in June) and an *Erax* (in Mariposa County). In May, June, and the beginning of July, I often visited localities in Marin and Sonoma Counties, also in the immediate surroundings of San Francisco (Lone Mountain for instance), where I could expect to find *Erax, Promachus, Proctacanthus*, without finding a single specimen, while *Dasypogonina* were abundant.   In the Sierra Nevada, in July, I found an *Erax*, a *Machimus*, a *Tolmerus*, and a species respecting the position of which I am in doubt, and which occurred quite abundantly about Webber Lake.

The most striking peculiarities of the Californian fauna, as far as known, consist in the occurrence of several genera of *Dasypogonina*, hitherto not found anywhere else (*Ablautatus, Dicolonus, Callinicus*), and in the great abundance of species of *Cyrtopogon*, especially in the higher regions of the Sierra Nevada.   In one locality, Webber Lake, I found thirteen species, nearly all on the same day, a number which exceeds that of all the known European species.   *Daulopogon* also is well represented.

The genus *Nicocles*, hitherto found only in North America, is represented by two species in the Atlantic States and two in California; one occurs in Mexico.

Clavator, if my identification be correct, is common to California and Chili.

Among the genera characteristic of the western region in general, I will name the following:—

Ospriocerus, represented now by four or five species, occurs everywhere from Texas to California; also in Mexico.   It is not known to occur outside of North America.

Stenopogon, with ten described and many undescribed western species, *Scleropogon*, with two described species, and *Saropogon*, with two species, from Texas, also occur abundantly in the countries adjacent to the Mediterranean, the Black, and the Caspian Seas, and extend into Central Asia.

Microstylum, with two or three species in Texas and Kansas, is very abundantly represented in Southern Africa, and occurs also in the East Indies and Australia.

LAPHRIA.

1. LAPHRIA (DASYLLIS) ASTUR n. sp., ♂ ♀.—Like *Laphria posticata* Say, but the tibiæ beset with yellow pile. Length 14–20^{mm}.

Black; face, occiput, thoracic dorsum, and the two penultimate abdominal segments densely clothed with yellow hairs; palpi with black pile; a tuft of black bristles above the mouth is usually concealed under the overhanging yellow hairs of the face; scutellum with black pile. Legs black; front femora on their hind side and all the tibiæ with long yellow hairs; on the hind tibiæ, the yellow hairs do not quite reach the tip. Proximal half of the wings subhyaline; distal half more or less brownish, the inside of the cells being paler.

Hab.—Common in California; Petaluma, April 27; Mendocino, April 29 (J. Behrens); Saucelito, May 16. Most of my specimens, however, I caught about Webber Lake, Sierra Nevada (July 23–25). I have six males and nine females.

This species varies in the extent of the yellow pile, especially around the neck, on the pleuræ, and on the legs. As a rule, specimens taken at high altitudes have more yellow pile than those taken at lower ones. A specimen taken at Petaluma, therefore, but little above sea level, had no trace of yellow hairs on the tibiæ. The specimens from Webber Lake had a great many yellow hairs. The fan-like row of hair in front of the halteres is, in different specimens, either yellow or black. Some specimens have a yellow tuft in front of the wings, and another in front of the coxæ; in others, they are wanting. On the tibiæ, the yellow hairs appear sometimes only at the base.

2. LAPHRIA (DASYLLIS) COLUMBICA Walker, The Naturalist in Vancouver's Island and British Columbia, by J. K. Lord, London, 1866, 338.

" *Male.*—Black, with a very slight bronzy tinge; head very thickly clothed with slightly gilded hairs; vertex and hind side with black hairs; mystax composed of black bristles. Thorax clothed with short black hairs; fore part with a fawn-colored pubescence; a band of slightly gilded hairs across the hind part of the scutum. Abdomen clothed towards the tip with slightly gilded hairs; legs mostly clothed with slightly gilded hairs, except towards the tips; hind femora incrassated, with black hairs; hind tibiæ livid, and with slightly gilded hairs, except towards the tips. Wings blackish, discs of most of the areolets cinereous; veins and halteres black. Length of the body 9 lines; of the wings 16 l.

" This species has most resemblance to *L. posticata*, from which it may be distinguished by the pale hairs on the hind tibiæ."

I owe to the kindness of Mr. Henry Edwards male and female specimens from Vancouver Island, which I refer to this species, although some of the statements in the description do not quite agree with them. It is very like *L. astur*, but it has a band of black pile across the middle of the thorax.

3. LAPHRIA VULTUR n. sp., ♂ ♀.—Whole body clothed with reddish-fulvous pile, especially dense on the abdomen; legs black; the black of the thorax as far as visible among the red pile has a bluish opalescent reflection. Length 15–20ᵐᵐ.

There is but little to add to this diagnosis. The abdomen, especially in the female, is much more slender than in *L. astur*. The front legs on their hind side, sometimes also the middle legs are clothed with reddish-fulvous pile; much more in the male than in the female. The wings are hyaline on the proximal, infuscated on the distal half. Male forceps beset with reddish-fulvous pile.

Hab.—The woods of the Coast Range, above Santa Cruz, Cal., May 22 ; Webber Lake, Sierra Nevada, July 27. I received a specimen from Oregon from Mr. H. Edwards; two males and one female.

4. LAPHRIA RAPAX n. sp., ♂.—Head, posterior part of the thorax, and two first abdominal segments with whitish, the remainder of the abdomen, except the genitals, with ardent rufous pile; legs black. Length 20ᵐᵐ.

The lower part of the head and the base of the proboscis beset with whitish pile; face likewise, but many black, erect hairs are mixed with the white ones; hair under the antennæ altogether black. Front part of the thoracic dorsum with short, black pile; the hind part with longer, semi-recumbent, whitish pile; scutellum with some whitish pile; male forceps very large; wings, as usual, brownish on the distal half and hyaline on the proximal.

Hab.—Webber Lake, Sierra Nevada, July 28. A single male.

LAMPRIA.

LAMPRIA FELIS n. sp., ♀.—Head, thorax, base of the abdomen, and coxæ black; the rest of the abdomen and the legs red; wings tinged with brownish. Length 11ᵐᵐ.

The pile on the head is altogether black, except a small tuft of silvery hairs on each side, above the mystax, near the eye. The black of the thoracic dorsum shows a bluish opalescence, a pair of small spots of silvery pollen anteriorly, and another pair of less distinct, similar spots on the humeri. Halteres yellow. First abdominal segment and a large semi-circle on the second bluish-black. Alula and the proximal part of the axillary cell hyaline.

Hab.—Webber Lake, Sierra Nevada, California, (July 26). A single female.

CERATURGUS.

CERATURGUS LOBICORNIS n. sp., ♂ ♀.—The third antennal joint as well as the first joint of the antennal style have the terminal lobes prolonged beyond the insertion of the following joint. Thorax black, beset with short, appressed, golden-yellow pile. Abdomen yellowish-red ; venter black. Legs reddish-yellow ; base of front and middle femora black. Length, male, 10–11ᵐᵐ ; female, 11–12ᵐᵐ.

Face and front black, shining; above the mouth a row of yellowish bristles ; a few similar bristles on each side of the front, near the eyes. Antennæ black ; third joint a little longer than the two first taken together ; at the end, with two lobes, projecting beyond the insertion of the next joint ; the two following joints, forming the so-called antennal style, taken together are somewhat shorter than the third joint ; the first of them is a little shorter than the second, and has two projecting lobes, longer than those of the third joint. Thorax black, clothed on the dorsum with moderately dense, short, appressed, golden-yellow hairs, under which the shining black surface is visible ; in the middle, a gemi-nate, nearly bare, black stripe, formed by lines of denser pile on its sides and in the middle ; on the anterior part of the pleuræ, a dense patch of golden pile, followed by a shining black spot behind. Halteres yellow. Abdomen yellowish-red above, smooth, shining ; the extreme lateral margins of the segments black (more distinctly so in the female) ; base somewhat darker in the male ; venter black. Legs reddish-yellow ; coxæ and trochanters black ; base of four anterior femora more or less black ; tarsal joints more or less brown at the tip ; the fifth, except the root, altogether brown. Wings in the male tinged with brown, especially near the anterior margin ; the apex subhyaline ; in the female, the basal half is yellowish, the posterior and distal portion brownish. (I suspect that the coloring of the wings is very variable.)

Hab.—Snake River, Idaho (C. Thomas). Two males and one female. I have a female from California (G. R. Crotch), the face and front of which are brownish-red. The thorax also shows traces of reddish about the humeri and on the pleuræ ; the venter is red, and the four anterior femora have no black at the base. I believe, nevertheless, that it is the same species. The antennæ of the specimen are broken. The specimens from Idaho had been kept in alcohol ; hence the antennæ are somewhat distorted.

DIOCTRIA.

Europe contains between twenty and twenty-five species of this genus while in the North American fauna it was hitherto represented by two rather aberrant species, *D. albius* Walker, from the Atlantic States, and *D. resplendens* Loew, from California. A third species, *D. pusio* n. sp., from California, is remarkably small, but nearer to the normal type of the genus than the other two.

1. DIOCTRIA ALBIUS Walker, List, etc., ii, 301.—I have several spe-

cimens from California (San Rafael, Marin County, May 29; Sonoma County, July 4) and Vancouver Island (G. R. Crotch), which are very like *D. albius* from the Atlantic States, but seem to differ in the abdomen being less narrow and less glabrous; the scattered, fine, yellowish-red pile of the abdomen, which is very little conspicuous in the Atlantic specimens, is very distinct in the Californian ones. However, *D. albius* from the Atlantic States is subject to variations the limit of which is, as yet, very doubtful to me. I have ten male and female specimens before me from the Catskill Mountains, New York, from the White Mountains, and from the Palisades, New Jersey, opposite New York. Some of the males have the whole axillary region of the wing distinctly whitish; in others, this whitish tinge is very distinct on the whole proximal half of the wing. One of the male specimens from the White Mountains has the proximal two-thirds of the wings pale yellowish, the distal third blackish, the tibiæ yellowish-red except at the tip, the front femora yellowish-red except a broad black stripe on their upper surface. This specimen may be a distinct species, but it is singular that the two specimens from Vancouver Island also have the base of the tibiæ yellowish-red; the femora, however, are altogether black. The other Californian specimens show the same tendency to vary in the coloring of the wings as the Atlantic specimens. In this uncertainty, I prefer not to describe my Californian specimens until a larger number can be procured.

2. DIOCTRIA RESPLENDENS Loew, Centur., x, 21.—California. I have seen (in Mr. Burgess's collection in Boston) a specimen of this easily recognizable species.

3. DIOCTRIA PUSIO n. sp., ♀.—Thorax and abdomen black; segments 3–5 of the latter, dull reddish; legs yellowish-red; hind tibiæ dark brown except the tip. Length 4.2^{mm}.

Antennæ long, black, inserted on a small protuberance; third joint (without the style) as long as the first two taken together; the style not quite half as long as the joint, with a small expansion at the base (somewhat like Meigen, tab. 19, fig. 20, only the style, in comparison to the joint, is longer in *D. pusio*); face with a pollen, which is golden-yellow above, silvery below; mystax of a few whitish hairs; front and occiput black; posterior orbits and two spots above the neck on the occiput silvery-pollinose. Thorax black, shining; dorsum with three faintly indicated lines of microscopic pubescence; the lateral ones expanded into triangles anteriorly; pleuræ with several spots with a partly silvery, partly golden reflection. Knob of halteres lemon-yellow. Legs red, including the coxæ; hind tibiæ dark brown, except the tip, which is red and somewhat incrassated; first joint of hind tarsi large and stout. Abdomen black, shining, smooth; second segment with a greenish reflection; the three following segments are reddish, but with darker, metallic reflections. Wings with a rather uniform, slight brownish tinge; anal cell open; veins brown, yellowish at base; venation normal.

Hab.—Sonoma County, California, July 4. A single specimen.

ABLAUTATUS.

A new genus, established by Dr. Loew (Centur., vii. 63) for a Californian species, under the name of *Ablautus;* modified later (Berl. Ent. Zeit., 1874, 377) in *Ablautatus.* The species, *A. trifarius,* is described from a female specimen.

I have a male and two females which undoubtedly belong to this genus, but apparently to a different species, as the legs are altogether black, and beset with white spines, including the tarsi. I caught my species in company with *Clavator sabulonum,* which seems to mimic it, as its body is almost exactly of the same coloring; both occur on sandy soil.

The genus is easily recognizable by its large ungues and the total absence of pulvilli.

The following generic characters belong to *A. mimus,* but seem in the main to agree with *A. trifarius.*

Front and *face* comparatively narrow ; face almost flat, with a dense mystax, reaching nearly up to the antennæ; front very little broader above.

Antennæ.—Third joint by about one-half longer than the two first taken together, elongated, with a coarctation a little before the middle, and a slight incrassation beyond the middle (not unlike the third joint in an ordinary *Cyrtopogon,* only both the contraction and expansion are more marked); antennal style short, less than one-fifth the length of the joint, cylindrical, with a microscopic bristle; the two basal joints of the antennæ have, on the under side, several conspicuous bristles, some of them more than half the length of the antennæ; some very small hairs on the upper side.

Thorax.—Besides the usual hairs and bristles, there are some conspicuous bristles on its front part, on the sides of the median stripe, which do not exist in *Cyrtopogon;* scutellum with a row of long, erect bristles on its edge.

Eyes with the facets of the middle region very much enlarged (much more so than in *Clavator*).

Abdomen rather narrow, moderately convex, flatter in the female than in the male, gently tapering toward the tip; male hypopygium rather small; ovipositor with the usual star of bristles. At the base, on the sides, the usual long hair, besides which, on the first segment, a fan-like row of bristles is perceptible, which are white in *A. mimus,* and are described as "lutescent" for *A. trifarius.* I believe I see similar, but shorter, spines on the following segments, but they are white, like the pubescence surrounding them, and difficult to distinguish from it. Besides the longer white pile on the under side, the sides of the abdomen and the hypopygium are clothed with short, more or less recumbent, white pile.

Legs of moderate length and stoutness, very hairy, and beset with bristles, not only on the tibiæ, but also on the upper side of the front and hind femora; ungues remarkably long, and no vestige of pulvilli; the two last joints of the front tarsi in *A. mimus* are ornamented with a dense brush of short bristles.

Wings like those of *Cyrtopogon;* all posterior cells open; anal cell with a narrow opening; small cross-vein a little beyond the middle of the discal cell.

1. ABLAUTATUS TRIFARIUS Loew, Centur., vii, 63, *female* (California).

2. ABLAUTATUS MIMUS n. sp., ♂ ♀.—General coloring brownish-gray, with a series of rounded blackish spots along the middle of the abdomen, one at the base of each segment; larger black spots in the anterior corners of the same segments form two lateral series; segments 7 and 8 in the male black, shining, beset with white pile; segment 7 in the female black, shining. Thorax gray, with the usual three darker stripes. Legs black, densely beset with rather long, recumbent white hairs, and long, white, erect bristles; ungues black. The two last joints of the front tarsi in the male appear incrassated, because they are densely beset with black and yellow, recumbent, and closely packed short bristles, forming a kind of brush, the end of which reaches considerably beyond the ungues; the under side of this brush is black, on its upper side it is mixed of black and yellow; the ends of the first three joints of the front tarsi are armed with strong bristles, or spines, which are black, with a yellowish root; a couple of such spines in the middle of the first joint. In the female, the front tarsi are simple, and all the spines upon them are white, like all the other spines on the legs. Halteres honey-yellow. Antennæ black, the spines on the under side of the first two joints brownish-yellow. Fan-like fringe of hair in front of the halteres white. Mystax white, a few black bristles above the mouth; occiput with white pile. Wings very hyaline; veins black. Length 7–8^{mm}.

Hab.—Crafton, near San Bernardino, Southern California, in March.

OSPRIOCERUS.

1. OSPRIOCERUS ÆACUS Wied., i, 390 (syn. *Dasypogon abdominalis* Say).—Not rare about Colorado Springs, Colo. (P. R. Uhler). Wiedemann's description agrees with the female; in the male, the sixth and seventh abdominal segments are red; the hypopygium black. I also have two males from Spanish Peaks, Colo., June 15 (W. L. Carpenter), which agree with the others.

2. OSPRIOCERUS EUTROPHUS Loew, Berl. Ent. Zeit., 1874, 355, *female* (Texas).—I have seen two females from Kansas (G. F. Gaumer), which belong here; one of them had the thoracic dorsum reddish, a variety which Dr. Loew mentions as occurring in *O. rhadamanthus.*

3. OSPRIOCERUS RHADAMANTHUS Loew., Centur., vii, 52, *male* (Pe-

cos River, New Mexico).—I have no specimen of this species. Is it not the male of the preceding? The difference in the coloring of the abdomen between both corresponds exactly to the sexual difference of the same kind in *O. œacus.*

4. OSPRIOCERUS MINOS n. sp., ♂.—Altogether black; wings blackish. Length 17–18ᵐᵐ.

The face is slightly grayish-pollinose; the abdomen is more cylindrical, less flattened than in the male of *O. œacus;* the last antennal joint seems a trifle longer. In other respects, the specimen looks like a unicolorous *œacus.*

Hab.—Golden City, Colo., July 3 (A. S. Packard).

5. OSPRIOCERUS ÆACIDES Loew, Centur., vii, 51, California.—I do not know it.

STENOPOGON and SCLEROPOGON.

These genera, especially the former, are very abundantly represented on the western plains and in California. Ten species of *Stenopogon* and two of *Scleropogon* are described by Dr. Loew from those regions. I have several species which, I believe, are new, but I would not attempt to describe them without comparing them with all the previously described species, several of which I do not possess.

DICOLONUS.

I do not know *D. simplex* Loew (Centur., vii, 56), the Californian species for which the genus was established.

CALLINICUS.

CALLINICUS CALCANEUS Loew (Centur., x, 32), for which the genus was established, is not rare in California. I found it about San Rafael, Marin County, May 27, and received several specimens from Mr. H. Edwards, also taken in Marin County.

CLAVATOR.

A species from Southern California agrees very well with the genus *Clavator,* established by Dr. Philippi for several species from Chili (Verh. zool.-bot. Ges., 1855, 699, tab. xxvi, f. 31). The first of these species, *C. punctipennis,* must be considered as the true representative of the genus; the other species have the third posterior cell closed and the antennæ of a different structure; hence it is very doubtful whether they belong in the same genus. The agreement of my species is with *C. punctipennis.*

This generic identification would be rendered certain, if it could be ascertained whether *Clavator punctipennis* belongs in the number of *Dasypogonina* which have a spur at the end of the front tibiæ. Dr. Philippi does not say anything about it, and may have easily overlooked this character. My *C. sabulonum* has such a spur.

Dr. Gerstaecker (Entom. Ber., 1865, 99 and 113) identifies *Clavator* with *Hypenetes* Loew, established for a species from Caffraria; unfortunately, he does say on what grounds this identification is based. Dr. Schiner (Die Wiedemann'schen Asiliden and Novara) reproduces this synonymy, without any remark. Now if *Clavator punctipennis* has, like my Californian species, spurs at the end of the front tibiæ, it cannot be the same thing as *Hypenetes*, which has no such spurs. An attentive scrutiny of Dr. Loew's description discloses other characters which I do not find in my specimen, but which it would be superfluous to discuss here.

Head not unlike that of *Cyrtopogon*, but much smaller and narrower; mouth comparatively much larger and broader, cut obliquely, so that in the profile the head below the face appears retreating; face short, excised, in the profile, under the antennæ, the gibbosity beginning immediately below; the mystax occupies the center of the gibbosity, without reaching the eyes or the antennæ; the front is not perceptibly broader above.

Antennæ.—First joint subcylindrical, short; the second still shorter; the third somewhat longer than the two first taken together, attenuated at the base for about one-quarter of its length and then expanded to three times the breadth of its narrow portion, then attenuating again toward the tip (the shape of the third joint holds the middle between the figures 1 and 2 on page 699 of Philippi); at the end, a minute cylindrical style, ending in a microscopic bristle.

Proboscis a little shorter than the vertical height of the head, directed downward; palpi rather long.

Thoracic dorsum on each side of the central stripe with a longitudinal row of long, stiff, erect bristles; there are seven or eight in each row; a number of similar bristles on the sides of the dorsum. I perceive two on the antescutellar tubercle, two others in front of these, and again two (sometimes three) above the root of the wing, in front of the suture; scutellum with six similar bristles.

Abdomen subcylindrical, narrow, somewhat broader at the base; male hypopygium not stouter than the abdomen; female ovipositor with a star of short spines.

Legs rather strong; tibiæ and tarsi spinous; front and middle femora with a single spine on the hind side a short distance before the tip; front tibiæ with an S-shaped spur at the tip; ungues long; pulvilli also.

Wings like those of *Cyrtopogon;* anal cell very little open, sometimes closed; small cross-vein about the middle of the discal cell; second submarginal cell considerably longer than the second posterior; all posterior cells open; fourth posterior slightly coarctate. (Compare Dr. Philippi's above quoted figure.)

CLAVATOR SABULONUM n. sp., ♂ ♀.—Yellowish-gray; thorax with a geminate blackish stripe; abdomen with a longitudinal row of blackish spots; wings hyaline. Length 7–7.5^{mm}.

Yellowish-gray; face whitish, with a tuft of white pile on the gibbosity; in the female with a few (I count six) black bristles above the mouth, which I do not perceive in the male; ocellar tubercle, in the female, with a tuft of stiff, black bristles (I count eight); in the male, these bristles are white, and the front shows on each side a row of similar, but smaller, white bristles; in the female, the latter bristles are very thin and small. Antennæ black; first joint with white pile beneath; second joint on the under side with a couple of black bristles; occiput with yellowish bristles above, and with long, soft, white hairs below. Thoracic dorsum with a geminate brown stripe in the middle, and two broader stripes on the sides, abbreviated long before the humeri; the fan-shaped fringe of pile in front of the yellow halteres is white in the male, black in the female. Abdomen yellowish-gray, with whitish-gray reflections; an ill-defined, elongated, darker spot, not reaching the posterior margin, in the middle of each segment; a similar dark spot on each side of the segments 2-6; the last segment in the female shining brownish-black; in the male, hypopygium black, shining, with long white pile. Wings hyaline, a little less pure hyaline in the female, in which a strong lens shows hardly perceptible vestiges of brown clouds on the cross-veins. Legs black; femora at the base and tip and base of tibiæ red; tarsi brownish; the spines on the tibiæ in the male are mostly white; some black spines are perceptible on the upper side, especially of the front tibiæ; in the female, the spines are black; very few white ones are visible.

Hab.—Crafton, near San Bernardino, Cal., March, on dry, gravelly soil. Two males and one female.

PYCNOPOGON.

I have never seen a specimen of this genus, and have to rely on the statements of Dr. Loew (Linn. Ent., ii, 526). These statements convince me that I have a species of this genus before me, or at least one closely allied to it. The characters of the species are so well marked that it will easily be recognizable.

PYCNOPOGON CIRRHATUS n. sp., ♂.—Black; thorax with white hairs; abdomen with recumbent, golden-yellow pile, especially dense on its latter part; femora black; tibiæ red; middle tibiæ before the middle with a tuft of black pile. Length 8.5^{mm}.

Head and face clothed with white pile; some black bristles above the mouth and also in the upper part of the occiput. Thorax black (the dorsum is greasy in my specimen), with long, soft, white pile; the usual bristles black. Halteres lemon-yellow; the fan-shaped tuft in front of them rather dense, pale yellow. Abdomen black, shining, finely and sparsely punctate; segments, beginning with the second, clothed with recumbent, silky golden-yellow hair, growing gradually more dense on each subsequent segment; this hair is less dense at the bases of segments 2-5; sides and under side beset with long, yellow hair. Femora black, with white pile; tibiæ red, with white pile and black and white bristles; the middle tibiæ, on the front side, before the middle, are ornamented with a conspicuous tuft of black bristles, projecting on each side; hind tibiæ with a brownish ring a little before the middle. Wings feebly tinged with yellowish-brown on the proximal two-thirds, the rest hyaline; fourth posterior cell coarctate; anal cell slightly open.

Hab.—Foot-hills of Mariposa County, on the road to Clark's Ranch, beginning of June. A single male.

CYRTOPOGON.

The large number of species of this genus occurring in California is very remarkable. While only ten or eleven species are known from the whole of Europe, I found thirteen species, eleven of which were new, almost all on the same day, near Webber Lake, Sierra Nevada. Another remarkable fact is the peculiar sexual ornamentation of some of these species, especially of the legs of the male, which, as far as I am aware, does not occur in Europe. One of the species so ornamented occurs in the Rocky Mountains.

The structural characters of these numerous species (several of which I left undescribed for want of good specimens) offer a great variety, and will facilitate a subdivision of the genus. The synoptical table I give is a very imperfect attempt at such a grouping. The two last species, *C. cerussatus* and *C. nebulo*, especially the latter, are only provisionally placed in the genus, for want of a better place. In using the analytical table which follows, attention should be paid to the sex of the specimen, as some of the species were described in one sex only.

Synoptical table of the species.

I. Scutellum rather convex above, its posterior edge rounded; surface black, sometimes pollinose at the base only, and beset with long pile; legs densely pilose, and rather stout (except in *C. princeps*, where they are more slender): ·

  A. Hind tibiæ more or less reddish in the middle:

    (*a*) Front tarsi ornamented in the male with dense silvery pile; middle tarsi with disks of black bristles:

      1. *callipedilus* ♂ ♀ ; 2. *cymbalista* ♂ ♀ ; 3. *plausor* ♂ ♀.

    (*aa*) Front and middle tarsi not ornamented as above:

      4. *montanus* ♂ ♀ ; 5. *leucozonus* ♀ ; 6. *aurifex* ♂ ♀.

  AA. All the tibiæ entirely black:

      7. *princeps* ♂ ; 8. *cretaceus* ♀ ; 9. *longimanus* ♂ ♀.

II. Scutellum more or less flattened above; surface densely grayish-pollinose; legs moderately hairy, and not very stout (rather hairy in *C. rattus*):

B. Abdomen black, with white spots or cross-bands on the hind
margins of the segments :

(*b*) Legs reddish :
  10. *profusus* n. sp. ♂ ♀.
(*bb*) Legs black :
  11. *evidens* n. sp. ♂ ♀ ; 12. *rejectus* n. sp. ♀ ; 13. *nugator*
  n. sp. (♂ ?) ♀; 14. *positivus* n. sp. ♂ ; 15. *sudator* n. sp. ♂ ♀.
BB. Abdomen gray, with black spots :
  16. *rattus* n. sp. ♂ ♀.

Aberrant species.

17. *cerussatus* n. sp. ♂ ♀ ; 18. *nebulo* n. sp. ♀.

Analytical table for determining the species.

(1) { Antennal style fully as long as the third antennal joint ; wings with
distinct brown clouds on the cross-veins and on the bifurcation
of the third vein ....18. *nebulo* n. sp., ♀.
Antennal style much shorter than the third antennal joint ; wings
without distinct brown clouds, etc., (2):

(2) { Third antennal joint red, (16):
Antennæ altogether black, (3):

(3) { Hind tibiæ more or less red in the middle, (4):
Hind tibiæ black, (9):

(4) { *Male :* front tarsi with a conspicuous dense fringe of silvery pile ;
two last joints of middle tarsi with a disk of black bristles ;
female : hind margins of abdominal segments white on the sides
only, sometimes very little, (7):
Male : front and middle tarsi plain ; *female :* hind margins of
abdominal segments 2–5 altogether white, (5):

(5) { Scutellum convex, black ; the brownish pollen at the base, if any, is
hardly perceptible ; densely pilose with long, erect pile ; face in
the middle with white pile, (6):
Scutellum flat, covered with dense gray pollen : longer hairs along
the edge only ; face altogether with black hairs,
10. *profusus* n. sp. ♂ ♀.

(6) { The fan-like row of hairs in front of the halteres is black,
4. *montanus* Lw., ♂ ♀.
The fan-like row of hairs, etc., is white ....5. *leucozonus*, Lw., ♀.

(7) { Scutellum shining black, (8):
Scutellum with gray pollen ....3. *plausor* n. sp., ♂ ♀.

(8) { Abdomen with white pile on the sides ..1. *callipedilus* Lw., ♂ ♀.
Abdomen with black pile on the sides from the very base,
2. *cymbalista* n. sp., ♂ ♀

(9) { Abdomen gray, with black, shining spots..16. *rattus* n. sp., ♂ ♀.
Abdomen black, with white cross-bands, (10):

(10) { White cross-bands on the anterior margins of the segments,
17. *cerussatus* n. sp., ♂ ♀.
White cross-bands on the posterior margins of the segments, (11):

(11) { Scutellum shining black; legs very hairy..9. *longimanus* Lw., ♂ ♀.
     { Scutellum with gray pollen; legs moderately hairy, (12):

(12) { The fan-like fringe of hairs in front of the halteres is white, (13):
     { The fan-like fringe, etc., is black, (15):

(13) { First abdominal segment with a white cross-band, occupying
     {   nearly the whole posterior margin......11. *evidens* n. sp., ♂ ♀.
     { First abdominal segment with white spots on the sides only, (14):

(14) { Wings brownish-hyaline on the distal half; ungues black, reddish
     {   at the base only; length 9–10ᵐᵐ..........12. *rejectus* n. sp., ♀.
     { Wings almost uniformly hyaline; ungues yellowish-white, black
     {   at the ends only; length 7–8ᵐᵐ......13. *nugator* n. sp., (♂?) ♀.

(15) { Front and face broad, with a hoary bloom..15. *sudator* n. sp., ♂ ♀.
     { Front and face rather narrow, the bloom upon them not hoary,
     {   14. *positivus* n. sp., ♂.

(16) { Tibiæ black, (17):
     { Tibiæ, except the tip, red................6. *aurifex* n. sp., ♂ ♀.

(17) { Scutellum black, shining, (18):
     { Scutellum with gray pollen..............8. *cretaceus* n. sp., ♀.

(18) { Hairs on the face black, mixed with some white (compare *longi-*
     {   *manus* Lw.).
     { Hairs on the face yellowish; front tarsi of the male unusually
     {   long; front and hind tarsi with silvery hairs on the upper side,
     {   7. *princeps* n. sp., ♂.

1. CYRTOPOGON CALLIPEDILUS Loew, Berl. Ent. Zeit.,1874,358, ♂ ♀.
—*Male.*—Black; thoracic dorsum with a very weak brownish pollen, forming an indistinct geminate stripe in the middle and some ill-defined
marks on the sides; long white pile on the face, the lower part of the
occiput, front part of the chest, fore coxæ, and the sides of the two first
abdominal segments; black pile on the remainder of the abdomen and
the top of the head; some scattered black hairs on the thorax above
and some black hairs in the mystax above the mouth. Femora black
with long, soft, white pile; tibiæ reddish, beset with blackish pile; tarsi
black, except the first joint of the four hind tarsi, which is often reddish
to a greater or less extent from the root; the front tarsi, beginning
with the second joint, are densely beset on their upper side with recumbent, short, silvery hairs, parted in the middle; the under side of the
same joints is densely beset on both sides with short, black bristles;
the two last joints of the middle tarsi have on each side a dense, flattened
tuft of black bristles, which form together a kind of disk, which is a
little broader than long. Wings grayish-hyaline, more hyaline on the
proximal half.

Female.—Head, and especially the face, covered with a dense whitish-
gray pollen; the thoracic dorsum covered with a brownish-gray pollen,
completely concealing the black ground-color, except at the four corners
and on the scutellum, which is black and shining; a geminate darker
line in the middle of the dorsum, not reaching the scutellum; pleuræ
likewise clothed with dense yellowish-gray pollen. Abdomen shining

black, the hind margins of the segments 2–5 with white triangles on each side. The hairs on head and chest are like those of the male, but of a less pure white; the white hair on the sides of the abdomen reaches to its tip, gradually becoming shorter. Legs like those of the male, but the sexual ornaments on the front and middle tarsi are wanting. Wings with the grayish tinge a little more saturate than in the male.

Length, ♂ 11–12^{mm}; ♀ 11–13^{mm}.

Hab.—Yosemite Valley, California, June 5–12; Summit Station, Sierra Nevada, July 17; Webber Lake, Sierra County, California, July 22–26. Eight males and seven females.

Dr. Loew (l. c.) has given a very detailed description of the male; that of the female must have been taken from a very imperfect specimen, as it is not recognizable.

2. CYRTOPOGON CYMBALISTA n. sp., ♂ ♀.—*Male.*—Like the preceding in the ornamentation of the four anterior tarsi and in the general coloring of the body, but with the following differences:—

The abdomen is uniformly clothed with black pile. The white pile on the lower part of the occiput and on chest and front coxæ is less long and conspicuous. The black pile on the upper part of the occiput reaches much lower here. Only the four anterior femora have some white pile on their posterior side. The brownish pollen on the thorax is hardly perceptible here. The front tibiæ and the tips of the four hind tibiæ are black. Besides the silvery hair on joints 2–5 of the front tarsi, some silvery pile is also perceptible on the first joint. The under side of the same tarsal joints is not beset with dense brushes of short black bristles, as it is in *C. callipedilus,* so that the white silvery hairs are visible from below, which they are not in the other species. The pulvilli of the four hind tarsi are brown, while in *C. callipedilus* they are whitish. The wings are a little shorter. The abdomen is slightly tapering from base to tip, instead of being nearly cylindrical, as in *C. callipedilus.* A vestige of a spot of whitish pollen is generally visible in the hind corner of the fourth segment.

Female.—Black, shining; thoracic dorsum with a slight brownish pollen, which is a little denser than in the male, but much less dense than in *C. callipedilus* ♀. The hair on the face is deep black; a little whitish pile on the lower part of the occiput and on the front coxæ; pile on the legs black; their coloring the same as in the male; only the front tibiæ sometimes are reddish at the base and along their front side; abdomen with small triangles of whitish pollen on the hind corners of segments 2–4, the largest on the fourth segment. The shape of the abdomen is different from that of *C. callipedilus* ♀; gradually tapering from base to tip, instead of slightly expanding about the middle. Length, ♂ 11–12^{mm}; ♀ 12–13^{mm}.

Hab.—Summit Station, Sierra Nevada, July 17; Webber Lake, July 23–24; both sexes found in each locality. Three males and four females.

3. CYRTOPOGON PLAUSOR n. sp., ♂ ♀.—Very like the two preceding

species in the ornamentation of the four anterior tarsi of the male, and at the same time very different, even in those characters.

Male.—Pile on the face pale yellow, sometimes yellowish-white, black above the mouth; lower part of the head posteriorly and front part of the chest with yellowish-white pile; thoracic dorsum, including even the base of the scutellum, covered with a yellowish-brown pollen, except at the four corners, which are black; three stripes on the dorsum are less pollinose, and therefore darker; the intermediate one geminate, and abbreviated posteriorly; the lateral ones broad, abbreviated, and rounded anteriorly, converging toward each other posteriorly, in front of the scutellum; abdomen black, shining, clothed on the sides with dense yellowish pile, gradually diminishing in length toward the tip. Legs black; tibiæ reddish, black at tip; front tibiæ often altogether black. Joints 2–5 of the front tarsi densely beset with silvery-white recumbent hairs along the outer and upper side only, and therefore not parted in the middle (in the two preceding species, the silvery hairs are found both on the outer and inner side of the upper part of the tarsi, and are parted in the middle); some silvery pile on the outside of the first joint; the two last joints of the middle tarsi with a disk-shaped, flat brush of black bristles, as in the two preceding species; pulvilli blackish-brown; all the femora and the four posterior tibiæ beset with long pale yellowish pile. Wings grayish-hyaline.

Female.—Like the male, except in the sexual ornamentation of the front and middle tarsi, etc. Abdomen black, shining, the sides densely beset with pale yellowish-white pile, through which, on segments 2–5, the white pollinose spots in the hind corners of the segments are visible. Will be easily distinguished from *C. callipedilus* ♀ by its pollinose scutellum, less densely pollinose thoracic dorsum, more yellowish pile of the face and chest, etc. Length, ♂ ♀ 12–13^{mm}.

Hab.—Morino Valley, New Mexico, July 1 (Lieut. W. L. Carpenter); Spanish Peaks, June (the same); Cache Valley, Utah (C. Thomas); divide between Idaho and Montana. Six males and two females.

4. CYRTOPOGON MONTANUS Loew, Berl. Ent. Zeitschr., 1874, 362.

"*Male.*—Ater, tibiis tamen posticis præter basim et apicem rufis, pilis nigris et albis vestitus, abdomine toto nigro-piloso, segmentis 2–5 singulis postice fascia albo-pollinosa ornatis, alis cinereo-hyalinis. Long. corp. 3¾ lin.; long. al. 2¾ lin.

(Translation.)—"Deep black; hind tibiæ dark reddish, brownish-black at base and tip; the upper half of the occiput and the front with long black hairs. The two first antennal joints with scarce, at least in part whitish hairs; third joint wanting. The dense mystax, reaching up to the antennæ, is black on the sides and on the lower part of the face; its inner part is white. Palpi with black hairs; lower half of the occiput and the mentum with white hairs. The thoracic dorsum seems to have been principally covered with grayish pollen; the specimen is too badly preserved to warrant a positive statement. The hairs on thoracic dorsum, scutellum, and pleuræ are black; some few scattered pale colored small hairs are mixed with them. Segments 2–5 of the shining black abdomen have each, on their hind margin, a moderately broad cross-band of whitish pollen. The cross-bands on segments 2–4 are interrupted in the described specimen (probably rubbed off?); that of the fifth segment seems to have been interrupted even in the intact specimen. The hairs on the abdomen and hypopygium are, without exception, black; a part of the hairs near the lateral margins has the ends of a lighter color. The hairs on the coxæ are whitish, those on the hind coxæ are mixed with numerous black hairs. The legs are of the same structure as in *C. longimanus* Loew, especially the front tarsi of the same conspicuous length; the pile on the femora is prevailingly, but not altogether, black; some whitish hairs on the under side, at the basis of the front femora, and short white hairs on the greater part of the upper side of the hind femora, are especially perceptible; the front tibiæ likewise have principally black hairs; on the hind tibiæ only the under side is beset with long black hairs, while elsewhere short white pile, on the upper side longer white pile, is prevailing. The tarsi are almost exclusively beset with black hairs. The bristles on the legs are without exception black (the halteres are wanting). Wings grayish-hyaline, hardly more grayish on the distal half; the veins normal, brownish black; the central cross-veins with vestiges of darker shades in their surroundings, which will probably not be visible in fresher specimens.

"*Hab.*—Sierra Nevada (H. Edwards)."

I have three males from Webber Lake, Sierra Nevada (July 22), which seem to belong to this species. The white pile on the face prevails over the black, which is distinct on the under side only, but very scarce on the sides; a very thin brownish pollen conceals but very little the black, shining thoracic dorsum; a median geminate stripe is hardly visible, but on the side of it a semblance of the figure 5 in gray pollen, with its reverse on the other side, is more distinct; the cross-bands of white pollen on the abdominal segments are all interrupted. Length 8.5–9.5^{mm}.

I will add the description of the females, taken in the same locality, which may belong here.

Female.—Thorax more densely brownish-pollinose; the double figure 5 on each side of the brown median stripe grayish-pollinose; the white abdominal cross-bands entire, except that on the fifth segment; sides of the abdomen beset with white pile, alternating with tufts of black pile at the base of the segments; in some specimens, however, these black tufts are hardly perceptible. The fan-shaped row of hairs in front of the halteres is black here, as it is in the male specimens. Length 10–11^{mm}. Four specimens.

5. CYRTOPOGON LEUCOZONUS Loew, Berl. Ent. Zeitschr., 1874, 364.

Female.—Ater, tibiis posticis tamen et metatarso postico rufis, pilis nigris et albis vestitus, abdomine toto albo-piloso, segmentis 2–5 singulis postice fascia albo-pollinosa ornatis; alis cinereo-hyalini s. Long. corp. $4\frac{5}{12}$ lin.; long. al. $3\frac{7}{12}$ lin.

(Translation.)—"Deep black; hind tibiæ red, their extreme base brown, the tip hardly infuscated; first joint of the hind tarsi dark reddish, brown-ish-black toward the tip. Occiput near the vertex and on the upper part of the posterior orbit beset with black pile; below, the pile is white or whitish; front with long black pile. The hairs on the first two antennal joints are whitish, (third joint wanting). The dense mystax, which reaches the antennæ, is black, with a moderate number of white hairs on the inside of its upper half. Palpi with black pile. The thoracic dor-sum seems to have been clothed near the humeri with whitish-gray, else-where with brownish-gray pollen. (The condition of the specimen for-bids any positive statement.) The hair on the thoracic dorsum is prin-cipally whitish; from the middle, however, toward the anterior margin, the blackish hairs become gradually more numerous. The usual bristles near the lateral margin and above the root of the wings are black. Nu-merous whitish hairs are mixed with the long black pile on the scutel-lum; the hairs on the pleur are exclusively whitish. Segments 2–5 of the shining black abdomen each have on their hind margin a moderately broad cross-band of white pollen; that on segment 5 is interrupted in the described specimen (probably denuded). The hairs on the abdomen are without exception white, longer at the base, shorter and more scarce toward the end, on the last segments erect, in the usual manner. The pile on all the coxæ is whitish, without any admixture of black hairs. The hairs on the femora are prevailingly but not exclusively whitish; those on the front femora toward the tip are mostly black on the upper and the front side; on the hind femora, most of the black pile is at the end of their posterior side. The hairs on the front and middle tibiæ are mostly black; on the upper side of the middle tibiæ, numerous white hairs are mixed with them; on the hind tibiæ, the under side is beset with long black hairs, while the remaining pubescence is white; on the upper side, rather dense and moderately long. The hairs on the tarsi are exclusively black; the bristles on the legs are also black. Halteres whitish-yellow, with a brown stem; wings grayish-hyaline, hardly more grayish on the distal half; the venation normal; the veins brown-ish-black; the central cross-veins show in their immediate surroundings distinct traces of a darker shade, which are probably less distinct in very fresh specimens.

" *Hab.*—Sierra Nevada (H. Edwards).

" *Observation.—Cyrt. leucozonus* is so very different from *C. montanus,* especially in the color of the pubescence of the whole thorax, of the ab-domen, and of the femora, that I do not dare to take it for the other sex of that species, although in the structure of the legs and in the position of the bristles on them their agreement is such as usually occurs between the sexes of the same species."

I have five female specimens from Webber Lake, Sierra Nevada, July

22, and Yosemite Valley, June 8, which agree with this description, with the single exception that on the face there is more than a "moderate number" of white hairs. The upper part of the mystax may be called white, with a few rare black hairs on the sides. I have no male to match these females.

C. leucozonus and *montanus* seem to belong to a group of closely resembling species. I have several specimens, among them a male from Salt Lake, Utah, which closely resembles the specimens which I have identified with Dr. Loew's descriptions, but cannot possibly be considered as the same species. I am not absolutely certain of having identified those two species correctly; nor am I very confident that the specimens which I described as the female of *montanus* really belong there. In order to render my possible error harmless, I have purposely reproduced Dr. Loew's descriptions, and abstained from describing any species of my own belonging to this group.

6. CYRTOPOGON AURIFEX n. sp., ♂ ♀.

Male.—Abdomen narrow, tapering, black, shining, with some bluish reflections on the first segment and very distinct purplish reflections toward the tip before the hypopygium; first segment on the sides with tufts of black pile anteriorly and white pile posteriorly; the hind part of the second and the greatest part of the third and fourth segments are occupied each by a conspicuous broad fringe of long, erect, yellowish-fulvous fur, with narrow bare spaces between these fringes. The three following segments are covered with short, dense, erect, deep black hairs, forming a brush, especially conspicuous on the sides, and longer posteriorly before the hypopygium; the purplish, black, shining ground-color is almost covered up by this pile; hypopygium black, shining, with but little pile. Face and front with brownish-gray pollen; face with whitish pile above and black pile below; occiput with white pile below and black pile above and on the vertex; third antennal joint red; the style black. Thorax black, brownish-pollinose, especially about the humeri; a brown geminate stripe, with a paler, grayish-yellow dividing-line in the middle. Femora black; front tibiæ red at base, black on their distal half; the other tibiæ red, broadly black at tip; tarsi black, the base of the first joint and the extreme root of the following joints red; three first joints of the front tarsi with some white pile on the upper side. Wings brownish-hyaline; fourth posterior cell hardly coarctate at all. Length 8.2^{mm}.

Female.—Legs like those of the male; only the red on the tarsi occupies more space and the front tarsi have no silvery pile; the hairs on the face are more scarce and whitish; the abdomen comparatively narrow, shining, black; segments 2–5 each with a moderately broad cross-band of yellowish-white pollen near the posterior margin; segments 2–4 are sparsely clothed with yellowish-fulvous erect pile, not concealing at all the ground-color, and not forming the fringes of fur so conspicuous in the male; segments 5–7 are almost glabrous, some very scarce, short pile being only perceptible. The thorax is more densely pollinose than in the male; a grayish pollen forms a V-shaped figure posteriorly, the apex of which rests on the scutellum, the ends branch off on each side along the thoracic suture; the geminate grayish stripe is longitudinally divided by a more yellowish line; the usual brownish shadows in the humeral region. Length 9–10ᵐᵐ.

Hab.—Webber Lake, Sierra Nevada, California, July 22. A male and female; in excellent preservation. It will not be difficult to recognize this remarkable species.

7. CYRTOPOGON PRINCEPS n. sp., ♂.—Front tarsi remarkably long, once and three-quarters the length of the tibia; their whole upper side beset with a dense fringe of silvery pile; hind femora, tibiæ, and tarsi on the upper side with a similar, but broader, covering of silvery pile. Face and front with a brownish-yellow pollen; mystax pale yellow, black only above the mouth and on the lower part of the face; lower part of the occiput with white, upper part and vertex with black pile; third antennal joint red, rather long and slender, the style black. Thorax black, somewhat shining posteriorly, and somewhat brownish-pollinose, especially about the humeri; scutellum black; thoracic pile black. Abdomen black, shining, with black pile; segments 2–6 with yellowish-gray pollen on the hind margins; on the second segment, this pollen is visible on the sides only; on the third and fourth, it forms an interrupted cross-band; on the two following segments, this cross-band is broader and only subinterrupted by a deep emargination; the sixth segment is entirely covered with the gray pollen, except a small black triangle in front; hypopygium black, with black pile, and only a small fringe of minute yellowish hairs at the extreme end. Legs black, ornamented as described above; ungues whitish, with black tips. Wings uniformly blackish; veins normal; fourth posterior cell hardly coarctate at all. Length 10.5ᵐᵐ.

Hab.—Webber Lake, Sierra Nevada, California. A single male. A very remarkable species, easily recognizable by the blackish tinge of the wings and the ornamentation of the front and hind legs.

8. CYRTOPOGON CRETACEUS n. sp., ♀.—Thoracic dorsum rather evenly clothed with a grayish-white pollen, completely concealing the ground color; the coloring of this pollen is rather uniform, a geminate median stripe is hardly perceptible; ante-scutellar callosities black, shining; scutellum black, brownish-pollinose at the base; pleuræ with dense yellowish-gray pollen. Abdomen shining black; segments 2–5 each with a moderately broad cross-band of yellowish-white pollen on the hind margins. Face and front densely covered with yellowish-gray pollen; mystax white above, black below above the mouth; vertex and upper part of the occiput with black pile; lower part with long white hair. Third antennal joint red or reddish, the style black. Thorax with fine black erect pile on the front part, and with whitish pile on the back part of the dorsum; the base of the scutellum with whitish pile, the remainder with long black pile. On the pleuræ, the fan-like fringe of pile in front of the halteres is mixed of

white and black hairs; the subhumeral callosities and the lower part of the pleuræ are beset with white hairs; but, in front of the root of the wings, there is some black pile. The abdomen on the sides is beset with white pile; it is long and tuft-like at the base, but becomes rather rare beyond the third segment. Legs uniformly deep black, shining; they are much less stout than in *C. callipedilus* ♀; the tibiæ, especially the front pair, are more straight; front tarsi rather long; the pile and bristles on the four anterior legs are black, except some white pile on the under and hind side of the femora; the hind femora and tibiæ are beset with white pile, which is particularly dense on the upper side of the hind tibiæ; the bristles, as usual, are black; the first joint of the hind tarsi shows, in a reflected light, some short, white pile; otherwise the tarsi are uniformly black. Ungues whitish, with black tips. Wings grayish-hyaline; venation normal. Length 10.5^{mm}.

Hab.—Webber Lake, Sierra Nevada, California, July 22. Two females. A third specimen, from the same locality, has the third antennal joint much darker reddish-brown; the thoracic dorsum has distinct brown stripes, and is less whitish; the fan-like fringe of pile in front of the halteres consists of black hairs only, etc. I believe, nevertheless, that the specimen belongs to *C. cretaceus.*

This species, like *C. princeps*, has the ungues whitish, with black tips; both species were found in the same locality; they are too different, however, to be taken, without further evidence, for the sexes of the same species. The other species of *Cyrtopogon*, described in this paper, have the ungues black and more or less brownish or reddish at the base only. *C. profusus* and *nugator* are the only species which, in this respect, resemble the two above-mentioned ones.

9. CYRTOPOGON LONGIMANUS Loew, Berl. Ent. Zeitschr, 1874, 360;

"*Male.*—Totus ater, pilis nigris et albis vestitus, vittis dorsalibus thoracis tribus latissimis fusco-pollinosis, segmentis abdominalibus secundo, tertio, quarto et quinto singulis postice fascia lata albo-pollinosa ornatis, alarum dimidio basali hyalino, apicali nigricante.—Long. corp. 4$\frac{1}{12}$ lin. long. al. 3$\frac{5}{12}$ lin. (about 9^{mm} and 7.5^{mm}).

(Translation.) "The ground-color of the whole body is, without exception, shining black. The front with a long black pubescence, with which are mixed some whitish hairs, or such which appear whitish toward their tip. Antennæ black, the two first joints sparsely beset with black hairs, partly whitish toward their tips; the third joint very slender, strongly coarctate in the middle; terminal style slender, a little more than half as long as the joint. The long mystax reaches up to the antennæ, and is composed in the middle of hairs which are whitish, or black at their base only; the hairs on its outer side, all around, are exclusively black, so that, seen from the side, the mystax seems to consist entirely of black hairs; the long pile on the occiput is white; in the vicinity of the vertex and on the greatest part of the posterior orbit, it is black. The pollinose design on the thoracic dorsum resembles that of *C. marginalis* Lw. It consists of three broad stripes cov=

ered with dense brown pollen; the lateral ones are considerably abbreviated anteriorly; the intermediate one, seen from the front side, appears entire; seen from the hind side, it appears bisected by a broad black line; the region in front of the lateral stripes is covered with a thin white pollen, of which there is also a trace in the intervals between the middle stripe and the lateral ones. These intervals do not show the shining surface of the broad lateral margin of the thoracic dorsum, which is entirely free from pollen; the inner end of the thoracic suture, on each side, shows a small spot of more dense whitish pollen. The thoracic dorsum is beset with long black pile, which is rather scarce, except on the shining black sides of the dorsum, where it is a little more dense. Among this black pile, there is a shorter and more delicate white pubescence; it does not exist, however, on the shining black portions of the dorsum. The shining black scutellum is rather densely beset with long, exclusively black, pile. Pleuræ with a thin grayish pollen; their pubescence in front of the halteres and of the roots of the wings is altogether black; above the front coxæ, the stronger hairs are black, the more delicate pile whitish. Segments 2–4 of the shining black abdomen have on the hind margin a very broad cross-band of white pollen, which is even expanded in the middle; a similar cross-band on the fifth* segment is less broad, and a little interrupted in the described specimen (perhaps, in consequence of detrition). The pile on the abdomen is rather long, but becomes gradually shorter toward its end. On the five first segments, it is chiefly white; however, the sides of the three first segments (exclusive of their posterior corners) bear some black pile, which may show a trace of whitish reflection on the tips of the single hairs only. This black pile reaches down to the venter. From segment 6 to the much developed hypopygium the pile on the abdomen is altogether black. Coxæ with whitish pile. The black legs do not show any trace of lighter color; they have the ordinary structure. The front tarsi are comparatively long, equaled only by those of *C. marginalis* and *montanus*. The hairs on the legs are long, chiefly whitish on the femora. At their tip, however, and on the upper and hind side of the front femora, the hairs are more or less exclusively black. On the under side of all the femora, especially toward their base, the hairs have a pale yellowish tinge. The pubescence of the front tibiæ is chiefly black; but on their distal half there is a good deal of white pile. On the hind tibiæ, the hairs are prevailingly white, although there are many black hairs near the base, on the under side more than on the upper side. The hairs on the tarsi are chiefly white, on the upper side of the three first joints of the hind tarsi comparatively long, otherwise short. All the bristles on the legs are black. Halteres blackish-brown. Proximal half of the wings hyaline; distal half blackish-gray; venation normal; veins black.

"*Hab.*—San Francisco (H. Edwards)."

I have three males from San Rafael, Marin County, Cal., March 31,

---

* The original has fourth; evidently a mistake.

which agree perfectly with Mr. Loew's description, except that the grayish-pollinose cross-bands on segments 2-4 have a distinct black emargination in the middle, which is not mentioned. The third antennal joint is sometimes reddish; the antennal style is long, still I would not call it longer than half of the third joint. The thorax of my specimens shows, on each side of the median stripe, anteriorly, a short streak of whitish pollen. Length 4-10^{mm}.

I will supply a short description of the female, of which I have three specimens, taken on the same day with the males; a fourth is from Sonoma County, end of April.

Female.—The cross-bands of whitish pollen on segments 2-4 are nearly parallel, slightly narrower in the middle, especially on the fourth segment, where they are more expanded on the sides; on the fifth segment, the cross-band is interrupted in the middle. The pile on the sides of the abdomen is white, with the exception of a tuft at the base of the first segment on the sides, and of a smaller one on the sides of the second segment. The prevailing pubescence on the tarsi is black; the white hairs on the hind tarsi, which are conspicuous in the male, are wanting here. The brownish on the distal part of the wings is much less distinct than in the male, hardly perceptible. The antennæ of one of the specimens are somewhat reddish toward the end. Length 10-12^{mm}.

10. CYRTOPOGON PROFUSUS n. sp., ♂ ♀.—Thorax, including its sides and corners, and scutellum, densely clothed with gray pollen; three brown stripes on the dorsum; the intermediate one geminate, reaching from the anterior border to the scutellum (where it becomes almost black), longitudinally bisected by a gray line; the lateral stripes broad, abbreviated anteriorly, and bisected transversely by a gray line along the thoracic suture; the two halves thus produced are about equal in size, the anterior one being nearly round; both are dark brown on their inner side; the hair on the dorsum is black; a fringe of black hairs along the edge of the scutellum. Abdomen black, shining; posterior margins of segments 1-5 with a moderately broad cross-band of white pollen; in the male, the segments preceding the hypopygium are also whitish-pollinose; the sides of the abdomen on the basal half have tufts of long, soft, white hair; the fan-like fringe of hairs on the pleuræ in front of the halteres is white. Hypopygium beset with some black pile. Face and front brownish-gray, beset with black pile; in the female, I perceive some white hairs mixed with the black ones in the mystax. Antennæ black. Legs brownish-red; femora with black stripes along the upper side; they are beset with long, soft, white hairs; tibiæ with short white pile and black bristles, a few of the bristles on the middle and hind tibiæ are pale yellow; tarsi reddish-brown, almost black on the upper side; ungues whitish, with black tips; pulvilli whitish. Wings grayish-hyaline; venation normal. Length, male, 11-12^{mm}; female, 12-13^{mm}.

Hab.—Morino Valley, New Mexico, July 1 (W. L. Carpenter); Sangre de Cristo Mountains, July (the same). One male and three females.

The following five species have so many characters in common that a short statement of the distinctive characters of each will be more to the purpose than a long description. They were all taken near Webber Lake, Sierra County, California, July 22–24, nearly in the same locality. *C. rejectus* and *sudator* also occurred at Summit Station, Central Pacific Railroad, July 17. It is not improbable that several other species, belonging in the same group, exist in the same localities, or else that the species are subject to variation, as I have a few specimens left which I cannot refer to any of my species. The characters which these species have in common are : —

Face and front grayish- or brownish-pollinose, beset with black hairs ; occiput and mentum with white hairs ; antennæ black ; thorax grayish- or brownish-pollinose on the whole surface, with more or less well-marked darker stripes; scutellum densely grayish-pollinose, with a fringe of black hairs on the hind margin, and sometimes some pile on its upper surface; abdomen black, shining, with cross-bands of white pollen on the hind margins, sometimes entire, often more or less interrupted, sometimes so much as to leave white spots only on the extreme lateral ends of the margin ; this is always the case on the first segment, with the only exception of *C. evidens,* where the whole hind margin of the first segment is whitish-pollinose. The abdomen is beset on the sides of the segments nearer to the base with soft white hair. Halteres with a brownish stem and yellow knob; legs black, with black pile ; some white hairs on the under side of the femora, less numerous on the anterior femora. Wings hyaline on the proximal half, slightly tinged with grayish or brownish on the distal half (almost altogether hyaline in *C. nugator*). The fan-like fringe of hairs in front of the halteres is white in the first three, black in the last two, species.

In identifying these species, it must be borne in mind that all the specimens were taken at the same time and in the same locality, and that specimens taken at another season or in other localities may differ in the intensity of the coloring of the thoracic stripes or of the brownish tinge of the wings ; the white abdominal cross-bands may be also more subject to variation than I assumed them to be. Still, for each of the species, some permanent distinctive character will remain, as for *nugator* the shape of the abdomen, the hyaline wings, the color of the ungues; for *sudator*, the breadth of the front and its whitish, hoary pollen, etc. *C. rejectus* alone is doubtful to me, and may possibly be only a varity of *C. evidens.*

I. *The fan-like fringe of hairs in front of the halteres is white:*

11. CYRTOPOGON EVIDENS n. sp., ♂ ♀.—First abdominal segment with an uninterrupted white cross-band on the posterior margin ; stripes on the thorax very distinctly marked, brown ; the longitudinal dividing line of the geminate stripe is very distinct; the portion of the lateral stripe anterior to the thoracic suture is large, conspicuous, of a rich

dark brown; white cross-bands on abdominal segments entire, somewhat interrupted on the fifth segment only, rarely (in the male) on the fourth. Length, ♂ 7–8ᵐᵐ.; ♀ 10–11ᵐᵐ. Two males and four females. (In the male, the brownish tinge of the distal half of the wing is more marked here than in any of the four following species.)

12. CYRTOPOGON REJECTUS n. sp., ♀.—First abdominal segment whitish-pollinose on the sides only; the white cross-bands on segments 2–4 interrupted; on segment 5, the extreme sides only of the hind margin are white. Median germinate thoracic stripe less well marked and abbreviated earlier posteriorly; the portion of the lateral stripe anterior to the thoracic suture is well marked, brown. Length 9–10ᵐᵐ. Four females.

13. CYRTOPOGON NUGATOR n. sp., (♂?) ♀.—First abdominal segment whitish-pollinose on the sides only; the white cross-bands on segments 2–5 very markedly interrupted; length of the interruption nearly equal on segments 2–4. The abdomen (♀) is narrower, more cylindrical and convex, of more equal breadth from the base than in the ♀ of *evidens* and *rejectus*. Wings almost uniformly hyaline. Thoracic stripes very distinct, more blackish; the portion of the lateral ones preceding the suture is not conspicuously darker, and has not the rich dark brown color which distinguishes it in *evidens* and *rejectus*. Ungues whitish-yellow, the tips black. In size, this species is smaller, the ♀ being only 7–8ᵐᵐ. long. I have a male specimen which seems to belong here on account of its size and hyaline wings; but the stripe on the second abdominal segment is *not* interrupted (the thorax of the specimen is greasy); the ungues are whitish-yellow, with black tips, a very characteristic mark of the species.

II. *The fan-like fringe of hairs in front of the halteres is black :*

14. CYRTOPOGON POSITIVUS n. sp., ♂.—Peculiar coloring of the thoracic dorsum: the extreme anterior margin and the posterior beyond the thoracic suture are grayish; the intervening space is of a rich dark brown, the usual stripes coalescing completely; they reach for a very short distance beyond the thoracic suture; the longitudinal dividing line of the median stripe is very feebly marked with paler pollen (sometimes indistinct); front and face rather narrow, clothed with brownish pollen and dense deep black pile; first segment black, with very little white on the sides; posterior margin of abdominal segments 2–5 marked with white on the sides only, the interruption growing wider on each subsequent segment. Length, ♂ 7–8ᵐᵐ. Three males.

15. CYRTOPOGON SUDATOR n. sp., ♂ ♀.—Front and face distinctly broader than in the four preceding species, clothed with a whitish hoary bloom; pollen on the thorax also whitish-gray in most specimens; the brownish stripes are variable, but often feebly marked, although distinct. First abdominal segment marked with white on the sides only; in the female, the usual white cross-bands on segments 2–4 are entire, on the fifth interrupted, occasionally subinterrupted on the fourth; in

the male, interrupted on all segments. Length, ♂, 8.5–9^{mm}; ♀ 8.5–10^{mm}. Two males and eight females.

16. CYRTOPOGON RATTUS n. sp., ♂ ♀.—Altogether covered with yellowish-gray pollen; thorax with a geminate black line in the middle; abdominal segments on each side anteriorly with a shining black spot. Length 9–10^{mm}.

Face and front grayish-pollinose, the former with a white mystax; the bristles above the mouth are black; ocellar tubercle and occiput beset with whitish hairs; a small tuft of black hairs on each side of the vertex near the upper corner of the eye. Thorax gray, with a black double line in the middle, abbreviated before reaching the scutellum; the lateral stripes are paler brown, ill-defined, and crossed transversely by the grayish-pollinose thoracic suture. The pile on the thorax is whitish anteriorly, black posteriorly; bristles black. Scutellum gray, clothed with long, soft, whitish pile. The fan-like fringe of hairs in front of the halteres is white. Abdomen gray, clothed with soft, whitish pile; segments 2–6 on each side at the base with a large shining black spot, diminishing in size on each subsequent segment. In the female, these spots are much smaller. The seventh segment, in the female, is black, shining; the hypopygium of the male is also free of pollen, but beset with yellowish-white pile. Legs black, beset with long, white pile; bristles on the tibiæ also white, except the terminal ones and those on the front side of the front pair. Halteres with a lemon-yellow knob. Wings hyaline; veins black, normal.

Hab.—Webber Lake, Sierra County, California, July 22. Five males and one female. The antennal style is comparatively shorter here than in all the preceding species, somewhat coalescent with the third joint; the bristle at its tip is very distinct.

17. CYRTOPOGON CERUSSATUS n. sp., ♂ ♀.—Black; thorax whitish-pollinose; abdomen with white cross-bands on the *anterior* margin of segments 2–6; the sides of the same segments posteriorly each with a large white spot; wings hyaline; legs black, with white hairs. Length 6.5–8^{mm}.

Face covered with a white, hoary pollen; mystax black, more dense immediately above the mouth than higher up; facial gibbosity rather flat, little prominent; antennæ black, third joint three times the length of the two first taken together, narrow, almost linear; the style is very short, perhaps one-tenth of the length of the joint, cylindrical, with a minute bristle; ocellar tubercle rather large and broad, with deep grooves on each side between it and the orbit of the eye; both the tubercle and the opposite side of the groove are beset with black pile, which, commingling, form a distinct tuft on the top of the head; on each side of this tuft, along the orbit of the eye, there is a narrow margin of minute microscopic yellowish-white hairs; lower down, on the orbit, on a level with the antennæ, there is, on each side, a small tubercle, the upper and outer side of which is clothed with the sáme microscopic yellowish-white pile; the occiput is beset with white hairs, except in its upper part, where there are some black hairs. Thorax black, clothed with a thin, gray pollen; three indistinct stripes are somewhat brownish; the lateral ones are incurved and somewhat expanded ante-riorly, where they end in a brown spot above the humerus; the me-dian line is simple and rather indistinct; the dorsum is clothed with short, sparse, white pile and longer black bristles; some of the latter form two rows on the lateral thoracic stripes. Scutellum flat, with six black, conspicuous, erect bristles on its hind edge. The fan-like fringe of hairs in front of the halteres is usually mixed of black and white hairs, its upper part being black, the lower one showing some white hairs; in some specimens, principally males, it is altogether white. Abdomen black, shining, moderately convex, of nearly equal breadth; segments 2–6 anteriorly have a narrow cross-band of white pollen, not reaching the lateral margin; on that margin, in the posterior angles of each of the same segments, there is a large white spot. The two basal seg-ments have some long white hairs on the sides. Legs black, densely clothed with short appressed white pile, beset with longer white hairs and black bristles; hind tibiæ gradually incrassated from the base to the tip; first joint of hind tarsi also somewhat stout. Halteres pale brownish. Wings hyaline; venation normal.

Hab.—Los Guilucos, Sonoma County, July 5. Three males and five females.

This species differs in many respects from the typical ones of the genus. The broad ocellar tubercle with the deep grooves on each side, the peculiar tubercles near the eyes on each side of the antennæ, the row of erect bristles on the lateral stripes of the thoracic dorsum, the subclavate hind tibiæ, the shortness of the antennal style in proportion to the length of the third joint, the gently convex but hardly gibbose face, the conspicuous six bristles on the otherwise bare scutellum, are so many characters which are not found in the other species.

Half a dozen specimens, taken in Mariposa County and Yosemite Valley (June 3–13), resemble *C. cerussatus* in having the white stripes on the *anterior* margins of the thorax; but they have no tubercles near the eyes, and are abundantly distinct in many ways. The specimens being injured, I abstain from describing the species.

18. CYRTOPOGON NEBULO n. sp., ♀.—Gray; thorax with a geminate brown stripe; abdomen shining black, with white spots in the hind cor-ners of segments 1–5; wings with brown clouds on the cross-veins and on the bifurcation of the third vein. Length 8–9 ^{mm}.

Face and front grayish-pollinose, with black hairs; the hairs on the face, in a certain light, look whitish at the tip; occiput with white hairs. Antennæ black. Thorax grayish-pollinose, with brown stripes; the in-termediate one dark brown, geminate, abbreviated before reaching the scutellum, but coming in contact with a pair of elongated brown spots in front of the scutellum; scutellum convex, with rather dense, long, and soft white hair, and some blackish bristles along the hind edge;

pleuræ with whitish pile. Halteres with a dark brown knob. Abdomen black, shining, with white hair on the sides; segments 1–5 in the hind corners with a spot of white pollen of moderate size. Legs black, beset with long, white hairs; most of the spines are also whitish, especially toward the tip, the roots being often brownish. Wings hyaline, with black veins; central cross-veins, those at the distal end of the discal cell, the small cross-vein, and the bifurcation of the third vein are very distinctly clouded with brown.

Hab.—Webber Lake, Sierra County, California, July 22. One specimen.

This species does not properly belong in the genus *Cyrtopogon*, from which it differs in the shape of the antennæ; the third joint is gradually tapering from the base to the tip; the antennal style is quite as long as the third joint; altogether, the antennæ are like those of *Anisopogon* (*Heteropogon* olim); but the proportions of the body, the somewhat, although moderately, gibbous face, the character of the mystax, etc., are more like those of *Cyrtopogon*.

ANISOPOGON.

(*Heteropogon* olim, name changed in Berl. Ent. Zeitschr., 1874, 377.)

I have a species from California (G. R. Crotch) and apparently the same from Vancouver Island (H. Edwards); they are not unlike *A. gibbus* from the Atlantic States in stature, but certainly different, the wings being nearly hyaline. The specimens are not well preserved enough for a description.

HOLOPOGON.

A single female specimen, from Webber Lake, Sierra County, July 25, is nearly altogether black, and certainly different from the described species from the Atlantic States.

DAULOPOGON.

(Loew, Berl. Ent. Zeitschr., 1874, 377; formerly *Lasiopogon*.)

This genus seems to be quite abundantly represented in California. I have two species taken in the immediate vicinity of San Francisco, a larger one from Yosemite Valley, and two or three from Webber Lake, Sierra County, California. As the species of this genus are rather difficult to recognize from descriptions, I will describe only one, which has very marked characters.

1. DAULOPOGON BIVITTATUS Loew, Centur., vii, 57; additions in Berl. Ent. Zeitschr., 1874, 370.—I believe I recognize this species in some specimens which I took near San Francisco, March 28; only the small cross-vein is in the middle of the discal cell, rather than beyond it.

2. DAULOPOGON ARENICOLA n. sp., ♂ ♀.—Brownish-gray; abdominal segments 2–6 each with a pair of semicircular brown spots at the base. Length 7–8^{mm}.

Brownish-gray, sometimes with a tinge of yellowish; the mystax and the few hairs on the vertex and on the upper part of the occiput yellow-ish-white; those on the lower part of the occiput pure white; antennæ, black. Thorax with two, rather distant, brown stripes, expanded and somewhat diverging anteriorly; the hairs and spines on the thoracic dorsum are whitish; scutellum with a quantity of long, erect, whitish hairs on its edge; a semicircular impressed line parallel to this edge is very distinct. Abdominal segments 2–6 at the base each with a pair of semicircular brown spots, gradually diminishing in size on each subsequent segment; a vestige of such spots is also visible on the seventh segment. Hypopygium of the male black, beset with whitish pile and with an appressed tuft of yellow hairs above the forceps. In the female, the eighth segment is black, shining. Legs yellowish-gray, with short, appressed, whitish pile and yellowish-white bristles. Wings with a slight brownish tinge; small cross-vein before the middle of the discal cell; second posterior cell sometimes very narrow, in some specimens even petiolate; the fourth posterior, in some specimens, coarctate toward the end, even closed; these characters are very inconstant.

Hab.—San Francisco, Cal., on the sands about Lone Mountain, April 6, and again June 29. Four males and four females.

NICOCLES.

(Jaennicke, formerly *Pygostolus*, Loew, Centur., vii, 28.)

Of the two Californian species described by Dr. Loew, I have found only *N. dives*, at the Geysers, Sonoma County, in the first days of May. Mr. James Behrens had taken it a few days before near Mendocino.

Family DOLICHOPODIDÆ.

The first insect of this family which I found after my arrival in California was a *Hydrophorus* (Santa Barbara, January 28). Since then, until the middle of May, I did not come across a single Dolichopodid, although both the fauna and flora of the environs of San Francisco attain their fullest development from the end of March to about the middle of May, when the effect of the cessation of the rain begins to be visible. Among the exuberant insect life of that season I did not discover a single *Dolichopus* in sweeping the grass with my net, nor did I see a single *Psilopus* or *Chrysotus* running on the leaves of low shrubs (in Florida I used to catch them abundantly in such situations as early as March). May 14, I caught a *Hygroceleuthus* and a *Dolichopus*, one specimen of each, on the walls of the railroad station at San Rafael. Since then, in June and July, I found a few *Dolichopodidæ* along the streams of running water in Sonoma County.

All in all, I brought home, from Marin and Sonoma Counties, two *Hygroceleuthus*, one *Dolichopus*, two *Tachytrechus*, one *Liancalus*, and my new genus *Polymedon* ; one *Psilopus* from Yosemite Valley; from the High Sierra, two *Dolichopus*, one *Tachytrechus*, one *Scellus*, one *Hydro-*

phorus; sum-total, thirteen species. A new *Scellus* from British Colum. bia is also described here.

It is rather remarkable that two species of *Hygroceleuthus* should have been brought from the lower altitudes of California, together with only one species of *Dolichopus*, while only four other species of *Hygroceleuthus* * are known from the whole world, against more than a hundred *Dolichopus*. It is also remarkable that among the few *Dolichopus* from California described by Mr. Thomson, one should also be a *Hygroceleuthus*, perhaps identical with one of mine, but too insufficiently described, in a female specimen, for identification.

One of the *Tachytrechus* is apparently identical with a species from the Eastern States; the other one is closely allied to, but certainly different from, another species from the same region.

The most remarkable discovery in this family is the new genus *Polymedon*, with its extraordinary development of the face and of the cilia of the tegulæ.

It would seem that, on the whole, *Dolichopodidæ* are but poorly represented in California. The places to look for them, in the vicinity of San Francisco, are probably the marshes surrounding the bay, a locality which I have neglected to visit. In the Sierra, they are somewhat more abundant, in species as well as in specimens.

All the necessary information about the *Dolichopodidæ*, including the definition of the genera, will be found in Dr. Loew's work on them, in the Monographs of N. A. Diptera, vol. ii.

HYGROCELEUTHUS.

I refer to this genus two species, in which the face descends as far as the lower corner of the eye; moreover, the two basal joints of the antennæ, especially in one of these species, are considerably enlarged, a character which belongs to the typical *Hygroceleuthus* (compare Loew, Monographs, etc., ii, 17).

1. HYGROCELEUTHUS CRENATUS n. sp.—*Male.*—First antennal joint longer than usual, expanded on the inside so as to meet a corresponding expansion of the other antenna; second joint nearly as long as the first, on the inside of its basal half a similar expansion; both joints are black, except these enlargements on the inside, which are reddish-yellow; the first joint is beset on the outer and upper side with long and dense black pile; the second has some hairs on its end; third joint comparatively small, subtriangular, black; the dorsal arista appears unusually stout, from the dense pubescence which covers it. Face yellowish-white; cilia of the inferior orbit rather stout, golden-yellow. Thorax and abdomen bright metallic-green; tegulæ yellow, with yellow cilia, sometimes mixed with black hairs; lamellæ of the hypopygium yellow, with a narrow black border. Legs yellow, including the fore coxæ, which have a

---

* Two in Europe, one in Siberia, one in North America (Loew, Monogr., ii, 17).

greenish-black stripe on the outside; tarsi infuscated from the tip of the first joint; hind tibiæ slightly incrassated, on the inner side glabrous, and with a longitudinal furrow, the bottom of which is brownish. Wings grayish, slightly tinged with yellowish anteriorly, except the costal cell, which is subhyaline; costa with a stout swelling at the tip of the first vein; the posterior margin deeply indented at the end of the fourth vein.

Female.—The antennæ are a little smaller and less hairy, although they have the same structure and coloring; the face is broader; the hind tibiæ are not incrassate and not glabrous on the inner side; the costa is without swelling; the indentation of the posterior alar margin is present. Length 6–7^{mm}.

Hab.—Los Guilucos, Sonoma County, California, July 5. Two males and one female.

2. HYGROCELEUTHUS AFFLICTUS n. sp.—*Male.*—Similar in all respects to the preceding species, except that the antennæ are not much larger and not much more hairy than those of an ordinary *Dolichopus*; the first joint has the same yellow expansion on the inner side; the second joint is much smaller, and has only a vestige of yellow on the inner side; the pubescence of the arista is so fine as to require a much stronger magnifying power; the face is silvery-white; the cilia of the inferior orbit almost white; lamellæ of the hypopygium whitish. The hind tibiæ of the male have the same structure, only they are a little more incrassated and the shallow groove on their inside is broader and more distinctly tinged with brown; on each side of the second abdominal segment, there is a tuft of long yellow hair, which does not exist in my specimens of *H. crenatus*; the wings are the same as in the latter species. Length 6–7^{mm}.

Hab.—San Rafael, Cal., May 19. A single male.

Observation.—*Dolichopus lamellicornis* Thomson, Eugenies Resa, etc., 511, 114 (a female), judging from the description of the antennæ, must be a *Hygroceleuthus*. The description does not agree with my female of *H. crenatus*; but it may be the female of *H. afflictus*, which I do not possess. All the characteristic marks of the species do not, of course, exist in the female, which the author describes; but he does not even mention the indentation on the hind margin of the wing, which in *H. crenatus* at least exists in both sexes; and, if the same is the case with the unknown female of *H. afflictus*, this would exclude the synonymy of Mr. Thomson's species.

DOLICHOPUS.

The three species of *Dolichopus* before me, after comparison with the analytical table of the Eastern American species, in the Monographs, etc., vol. ii, 323, may be tabulated as follows:—

I. Prevailing color of the legs black; cilia of the inferior orbit black:
 1. *corax* n. sp.

II. Prevailing color of the legs yellow; cilia of the inferior orbit pale; tegulæ with black cilia; fourth longitudinal vein bent, but not broken:

    1. Antennæ altogether black; last joint of front tarsi in the male black, with a large lateral thumb-like projection; penultimate joint likewise enlarged, triangular; tip of hind tibiæ and hind tarsi black:

        2. *pollex* n. sp.

    2. Basal joints of the antennæ red; third joint red at base only, the rest black; front tarsi ♂ with three long and slender joints; fourth joint small, white; the last joint enlarged, lamelliform, black:

        3. *canaliculatus* Thoms.

Two more species are described by Mr. Thomson (Eugenies Resa, 512), *Dol. metatarsalis* ♂ ♀, and *Dol. aurifer* ♂ ♀. The descriptions are wanting in some very essential characters, as for instance the color of the cilia of the inferior orbit and of the tegulæ. Moreover, it is by no means certain whether these species belong to the genus *Dolichopus* in its present acceptation.

1. DOLICHOPUS CORAX n. sp.—*Male.*—Face dull yellow; cilia of the inferior orbit black; antennæ black. Thorax and abdomen of a rather dark metallic-green; pleuræ very little pruinose; tegulæ yellow, with black cilia; hypopygium rather large, black; lamellæ nearly black, yellowish-brown in the middle only. Legs black; front tarsi about once and a third the length of the tibiæ; last joint expanded into a large black lamel, which is fringed with short hair on the edge; hind tibiæ slightly incrassated, shining on the inside, but on their latter half with an opaque brownish streak. Fourth longitudinal vein very greatly bent; wings grayish, slightly tinged with brownish anteriorly.

Female.—Hind tibiæ not incrassated, and without the opaque brown streak; the other sexual marks also absent. Length 5–6^{mm}.

Hab.—Webber Lake, Sierra Nevada, California, July 24–25, common. Seven males and four females.

2. DOLICHOPUS POLLEX n. sp.—*Male.*—Face of a dull golden-yellow, narrower toward the mouth; cilia of the inferior orbit whitish; antennæ black, third joint rather pointed. Thorax and abdomen metallic-green, shining, sometimes coppery; hypopygium with rather large, whitish lamellæ, bordered with black; cilia of the tegulæ black. Coxæ black, with a whitish pollen. Legs reddish-yellow; front tarsi about once and a third the length of the tibiæ, whitish-yellow, the tips of the first three joints black; first joint more than one-half the length of the tibia; second joint about one-third as long as the first; third joint shorter than the second, slightly expanded toward the end; fourth joint nearly as long as the third, expanded, triangular, black; the fifth is also black and still more expanded, inverted heart-shaped, with one of the lobes much longer than the other, square at the end, thus form-

ing a stump-, or thumb-like appendage ; middle tarsi blackish, except at the base; hind femora with a few black bristles on the latter half of the under side, forming an incipient fringe ; hind tibiæ broadly black at the tip, glabrous on the inside; hind tarsi altogether black. Wings with a slight swelling of the costa at the end of the first vein ; fourth vein bent but not broken. Length 5–6ᵐᵐ.

Hab—Webber Lake, Sierra Nevada, California, July 24. Three males.

3. DOLICHOPUS CANALICULATUS (syn. *Dolichopus canaliculatus* Thomson, Eugenies Resa, 512.)—*Male.*—Bright metallic-green, with a slight yellowish pollen on the thorax. Face silvery, somewhat yellowish above ; antennæ red ; third joint brown, except the under side at the base ; cilia of the tegulæ black; lamellæ of the hypopygium unusually long, whitish, foliaceous, narrow at base, margined with brown ; the large emargination at the end has a smooth edge, not jagged, and beset with a few almost imperceptible hairs ; a smaller emargination alongside of the large one is beset with curved bristles, which extend along the inner edge of the lamella and are much less coarse than in other species. Legs, including front coxæ, pale yellow ; front tarsi about once and three-fourths the length of the tibia ; three first joints slender, stalk-like, nearly of equal length ; the fourth very minute, subtriangular, white ; the fifth lamelliform, black ; four posterior tarsi brownish, except at base; hind femora on the inner side with a fringe of long yellowish hairs ; hind tibiæ on the inside before the middle with a glabrous spot of a brownish-yellow, which sends out a glabrous line, running along the upper side of the tibia to very near its tip. Wings subhyaline ; third vein bent but not broken. Length about 5ᵐᵐ.

Hab.—San Rafael, Cal., May 14 and 27. Two males. I also have a female from Brooklyn, near San Francisco, July 11, apparently belonging here. The hairs on the hind femora of the male being on the inner side, and not along the lower edge, are somewhat difficult to perceive.

There can be but little doubt about the synonymy, although I do not quite understand the description of the hind tarsi; at any rate, the word " postici " is inadvertently omitted in that description.

TACHYTRECHUS.

Of the two Californian species which I possess, the one is identical with a species from the Atlantic States, the other closely resembles another species from the same region.

1. TACHYTRECHUS ANGUSTIPENNIS (syn. *Tachytrechus angustipennis* Loew, Monogr., etc. ii, 113, 3).—The specimens, males only, described by Mr. Loew, were from the District of Columbia. I have four males and one female from Los Guilucos, Sonoma County, July 6, and a single female from Summit Station, Central Pacific Railroad, in the Sierra Nevada, at 7,000 feet altitude. I hardly doubt of the specific identity. The only discrepancy from Mr. Loew's description is that the brown lines on the thorax are not " very much shortened",

but stop only a short distance before the scutellum. In the female, the three flattened bristles on the hind tibiæ do not exist, and the costa is only imperceptibly incrassate.

2. TACHYTRECHUS SANUS n. sp.—*Male.*—First antennal joint rather large, reddish-yellow, beset with black hairs on the upper side, especially long toward the tip; second joint small, placed on the under side of a projection of the first, yellowish; the third subtriangular, small, brownish, and yellowish on the inner side only; arista very slender, glabrous, half as long as the body, with a spatulate expansion at the tip, about one-third of which at the base is white, the rest black (the antenna is like that of *T. mœchus,* figured in Monogr., ii, tab. iii, f. 6, *d,* only the spatule at the end has none of the emargination represented on the figure, and is like tab. iv, f. 12, *c*). Face very long and narrow, somewhat broader below, golden-yellow, but without luster. Cilia of the posterior orbit black; thorax metallic-green, with two distinct bluish lines on the dorsum, which is very slightly grayish-pruinose, especially about the shoulders. Pleuræ somewhat yellowish-pruinose above the coxæ; abdomen metallic-green, whitish-pruinose. Cilia of the tegulæ black; hypopygium greenish, with a large patch of brownish-yellow velvety down near the root; lamellæ of moderate size, reddish-yellow, with black hairs on the apical and yellowish ones on the lateral edge. Prevailing color of the legs yellow; front coxæ of the same color, dusted with golden-yellow; their extreme root black; front femora with scattered black pile on the outer side; front tibiæ with a row of stiff, erect bristles on the inner side, and with another row of still longer bristles on the outside; both, when well preserved, extend beyond the middle of the tibia; front tarsi about the same length with the tibiæ, brownish from the end of the first joint; these tibiæ and tarsi, in a certain light, are silvery-pruinose; middle femora blackish at the extreme root; middle tarsi brownish from the tip of the first joint; proximal half of the hind femora greenish-black; knees infuscated; hind tibiæ black at tip; hind tarsi black; wings subhyaline, rather short. Length 5–6^{mm}.

Hab.—Webber Lake, Sierra County, Sierra Nevada, July 22. Two males.   *a. comh. 3 aa, food of June, Werneritia,*

T. sanus is very like *T. mœchus,* which I used to find abundantly at the Trenton Falls, New York; but the hind femora are broadly black at base, the hind tarsi altogether black, which color also invades the tip of the tibiæ; the antennal arista is much longer, the spatule at its end not emarginate; the thorax has two blue longitudinal lines, etc.

I have a female *Tachytrechus* from Sonoma County (July 6), which evidently belongs to a third species. It has yellow legs; tarsi black; tips of femora to a considerable extent likewise black; cilia of the posterior and inferior orbit whitish; thoracic dorsum yellowish-pollinose, with a median brownish-coppery stripe, which is not pollinose, and stops before reaching the scutellum; dorsal thoracic bristles inserted on brown dots, etc.

POLYMEDON nov. gen.

Face of the male prolonged downward, and dependent in the shape of a silvery sheet, or ribbon; in length, this ribbon is about equal to the upper part of the face between the antennæ and the lower end of the eye. In life, the ribbon is straight; in dried specimens, its end is usually bent inward.

Cilia of the very small tegulæ in the male unusually long (bent backward, they would almost reach the end of the second abdominal segment); they can be folded together like a fan, and then form a long tapering horn- or spine-like body. Those specimens which I examined in life had the cilia folded in that way; in the dried specimens, they are sometimes spread out.

These two extraordinary characters define the genus sufficiently; the other parts of the body may be described as follows:—

Antennæ comparatively short; first joint with a few hairs above; second small; third suboval, with a blunt end; arista subapical, of moderate length and stoutness (the antennæ resemble figure 10, *d*, on plate iv, of volume ii, of the Monographs, etc.).

Face of moderate breadth, nearly parallel, prolonged in the male, as described above; in the female, the face has the usual structure; its lower edge is nearly on a level with the lower corner of the eye, and is not straight, but somewhat angular (compare l. c., tab. iii, f. 6, *b*, where, however, the face is much narrower).

The usual frontal pair of diverging bristles is present; the other converging pair, inserted at the upper corners of the eyes, is so minute as to be almost imperceptible in the male; it has the usual size in the female.

Thorax.—In the male, the double row of minute bristles, usually present in the middle of the dorsum, is altogether wanting; it exists in the female; the lateral rows of larger bristles are present in both sexes.

Hypopygium comparatively large and stout, nearly sessile; external appendages lamelliform, rather small, fringed with bristles.

Legs of moderate length; first joint of the hind tarsi without bristles.

Wings.—The last section of the fourth vein turns off before the middle, and converges toward the third in a very gentle curve; fifth vein in the male obsolete before reaching the margin; in the female, very much attenuate. Costa (♂) so much swollen before the tip of the first vein as to fill out the whole costal cell, except small spaces at both ends; in the female, the swelling is hardly perceptible.

Polymedon is a mythological name.

POLYMEDON FLABELLIFER n. sp.—Face and front silvery-white in the male, as well as the dependent ribbon; greenish-white in the female; cilia of the inferior orbit white; palpi black; antennæ brown above, yellowish-red on the under side. The dark metallic-green thorax in the male is almost altogether covered with a very striking silvery bloom, especially on the dorsum; in the female, this bloom exists only on the pleuræ, the dorsum being metallic-green, with a coppery stripe in the middle. Abdomen metallic-blue in the male, slightly hoary, especially on the sides, and with black pile; more greenish and blackish in the female. Hypopygium, as well as its lamellæ, black. Tegulæ and their long cilia black. Legs black, only the knees yellowish; front tarsi somewhat flattened. Halteres dark brown. Wings tinged with grayish, more brownish anteriorly. Length about 5^{mm}.

Hab.—Los Guilucos, Sonoma County, California; not rare on stones in running waters; July 5, 1876. Four males and two females.

LIANCALUS.

In the Monographs, ii, 198, Mr. Loew divides the four known species of this genus into two groups. In the first, the appendages of the hypopygium are lamelliform, and the scutellum has only four bristles; in the second, the appendages are filiform, and the scutellum has six bristles. The Californian representative of the genus holds the middle between these two groups. It has lamelliform appendages of the hypopygium and six bristles on the scutellum.

LIANCALUS QUERULUS n. sp.—*Male.*—Metallic-green; thoracic dorsum bluish, with four coppery stripes; posterior orbit with long, soft, white hairs; antennæ black; cilia of the tegulæ pale yellow; halteres of a saturate yellow, brownish at the root; scutellum with six bristles. Legs dark metallic-green; tarsi black; front coxæ elongated, with long, soft, white hairs; first joint of front tarsi short; the second nearly three times as long, and appearing stouter in consequence of a dense brush of very short, microscopic, erect hairs along its whole inner side; third joint only a little longer than the first, with a similar brush of hairs; the two last joints taken together about equal to the third. Wings hyaline, with a large brown spot at the tip, limited anteriorly by the second vein, contiguous to the margin between the tips of the second and third veins, distant from it between the third and fourth veins; reaching but little beyond the latter; not sharply limited, rather evanescent on its proximal side. The metallic bluish-green abdomen is whitish-pruinose on the anterior half of each segment, somewhat blackish along the hind margins. Hypopygium with very small brownish lamellæ.

Female.—The front tarsi have not the same structure as in the male, the first joint being the longest; wings grayish-hyaline, without any apical spot, but with two small rounded brownish-gray spots, with evanescent outline in the first posterior cell, and a similar spot at the end of the discal cell. Length of ♂ and ♀ about 6^{mm}.

Hab.—The Geysers, Sonoma County, May 6.

SCELLUS.

1. SCELLUS VIGIL n. sp.—*Male.*—Thorax grayish above, with two approximate brown lines; abdomen and pleuræ copper-colored, partly

metallic-greenish ; wings subhyaline, with a double grayish spot on the great cross-vein, and a similar larger spot on the last section of the fourth vein ; anal appendages of the male narrow, white, blackish at the base. Length 3.5–4.5^{mm}.

Face brownish-yellow, narrow above, broader below ; antennæ black ; the ground-color of the front is concealed under a grayish pollen. Thorax above with a dense gray pollen, almost concealing the coppery ground-color ; two approximate brown lines in its middle stop some dis-tance before reaching the scutellum ; between their end and the scutel-lum, an opaque dark brown spot. Pleuræ coppery, with greenish re-flections, slightly pruinose. The scutellum, with two bristles, is green-ish, coppery, or purplish. Abdomen (very much shrunken and with-drawn in my specimens) coppery, pruinose above, brilliant coppery, and greenish on the sides. Anal appendages ribbon-like, white, blackish near the root. Legs metallic-green or coppery, with purple reflections ; tarsi black. The structure of the legs agrees in the main with the de-scription of the legs of *S. filifer* Loew (Monogr., ii, p. 210). Halteres whitish. Wings subhyaline, their root yellowish ; costa yellowish-brown before its junction with the first vein ; a double grayish spot on the great cross-vein, and a similar larger spot on the last section of the fourth vein ; the latter is well defined on the proximal, and evanescent on the distal side.

Hab.—Webber Lake, Sierra Nevada, July 22–24. Three males, found resting on stones on hill-sides. *Museum Dia. Comun.*

This species differs from *S. filifer* Loew (Fort Resolution, Hudson's Bay Territory) in the coloring of the wings, which have no longitudinal gray streaks between the veins, the color of the anal appendages, which are not yellow at the end, etc. Nevertheless, the resemblance between the two species must be very great.

2. SCELLUS MONSTROSUS n. sp.—*Male.*—Thorax brownish-gray, with several rows of brown dots, on which the bristles are inserted, and two approximate brown lines ; wings tinged with brownish ; anal append-ages of the male at least as long as the abdomen, white ; their end brownish-yellow, inverted-spoonshaped. Length 6–7^{mm} (without the appendages).

Face brownish-ocher-yellow ; antennæ black ; front dull greenish-gray ; inferior orbit beset with yellow hair ; the superior with stiff, black spines. Ground-color of the thorax concealed under a thick grayish-brown pollen ; three rows of brown dots, in linear groups of three or four, bear the usual dorsal bristles ; on each side of the inter-mediate row, there is an uninterrupted brown line, reaching to the scu-tellum ; the coppery ground-color of the thorax is visible on the dorsum above the wings ; a large, coppery, shining spot on the upper part of the pleuræ ; a smaller one at the foot of the halteres ; abdomen copper-colored ; halteres yellow, the extreme root brownish ; tegulæ with yel-low cilia. Anal appendages at least as long as the abdomen, ribbon-like, white, except at the root, which is brown; they are angularly bent in the middle, the latter half expanded, inverted-spoonshaped, yellowish brown, bearing a fan-shaped tuft of long hairs at the end. Legs metallic-coppery; tarsi black. Lobe at the end of the front tibiæ very large, deeply emarginate at the base; the long spine on the inner side of the tibiæ appears bifid, from a strong bristle near its tip; middle tibiæ, besides some stiff bristles on the upper and under side, with a fringe of soft hairs on the hind side, which become longer toward the tip, and end there in a tuft of curly hair; the hind tibiæ end in a very long curved spine, hook-shaped at the tip (if stretched out, it would be nearly as long as one-third of the first joint of the hind tarsi); a smaller spine near it. Wings yellowish at the root, otherwise tinged with brown, especially between the first and third veins; costal cell tinged with yellowish; a brown cloud on the great cross-vein; another on the curvature of the fourth vein; some subhyaline spots near the root of the wings, the most conspicuous of which on the proximal end of the third posterior cell.

Hab.—British Columbia (Crotch). A single male.

HYDROPHORUS.

Two species: one from Webber Lake, Sierra Nevada (July 25), is allied to *H. innotatus* Loew from Sitka (Monogr., ii, p. 212) in the coloring of the face, the upper part of which is greenish, and in other characters; its halteres, however, have a yellow, and not an infuscated knob. The other species, taken near Santa Barbara (January 25), is easily distinguished by the color of its first longitudinal vein, which is brownish-yellow; the costa beyond the junction with the first vein is of the same color. I abstain from describing these species, as my specimens are not numerous enough nor well preserved enough for that purpose.

Medeterus breviseta Thomson (Eugenies Resa, etc., 510) from California is a *Hydrophorus*, as the author compares it to *Medeterus litoreus* Fallen, which belongs to that genus.

PSILOPUS.

I found a single species abundant in Yosemite Valley about the beginning of June. It is closely allied to *Psilopus melampus* Loew (Monogr., ii, 253) from Mexico, but shows some differences, especially in the structure of the legs. It would not be safe to describe it without the comparison of specimens of *P. melampus*.

Family EMPIDÆ.

Is abundantly represented in California. I have nine species of *Empis*, taken in Southern California in February and March; in Marin and Sonoma Counties in April and May; and about Webber Lake, Sierra

County, in July. The males of most species are provided with very remarkable appendages on the hind legs.

Rhamphomyia is represented in my collection by more than a dozen species, taken in the same localities.

Of the group *Tachydromina*, I have two or three species, belonging to the genera *Platypalpus* (*Tachydromia* Meig.) and *Tachypeza*.

The only species hitherto described are :—

EMPIS BARBATA Loew, Centur., ii, 19.—California.

RHAMPHOMYIA LUCTUOSA Loew, Centur., vol. ii, 290 (syn. *R. lugens* Loew, Centur., ii, 30).—California.

Family LONCHOPTERIDÆ.

LONCHOPTERA sp.—San Rafael, May 29.

Family PLATYPEZIDÆ.

PLATYPEZA sp.—Santa Barbara, Cal., January 29.

Family SYRPHIDÆ.

As I left San Francisco, going east, about the middle of July—that is, before the best season for collecting *Syrphidæ*, which is August and September, had really begun—my collections cannot be expected fairly to represent the fauna.

The species of *Paragus* (1 species), *Pipiza* (1 sp.), *Orthoneura* (1 sp.), *Chilosia* (3 sp.), which I found, have a remarkable family resemblance to the species from the Atlantic States and to the European ones; some of them will probably prove to be identical. The same may be said of *Melanostoma* (3 sp.), *Syrphus* (6 sp.), *Mesograpta* (2 sp.), *Sphærophoria* (3 or 4 sp.), *Allograpta* (1 sp.), *Baccha* (2 sp.).

In the genus *Syrphus*, the common occurrence of the European *S. pyrastri* in California, Utah, and as far east as Colorado, is remarkable when we recollect that it has never been found east of the Mississippi. Macquart received it from Chili. The question how it got to these regions is an interesting problem.

The other species of *Syrphus* are either identical with eastern species or closely allied to them. Both *Mesograptæ* are species also occurring in the Atlantic States. The *Sphærophoriæ* seem to be more numerously represented in California than east of the Mississippi.

Peculiarly western is the new genus *Eupeodes*, with a single species of common occurrence from California to Colorado.

Of *Eristalis* I have three species from California, two of which reach eastward to Colorado.

Helophilus is represented by two species, one of which (*H. latifrons* Lw.), common in the environs of San Francisco, has a wide easterly range, to Nebraska and probably beyond. The other (*H. polygrammus*) is a very peculiar form, and occurs high up in the Sierra Nevada and in Oregon.

In the genus *Volucella*, I have *V. mexicana* Macq. from Southern California and a new species. I also describe a *Volucella* and a *Temnocera* from Colorado.

Mallota posticata (with some doubt), *Polydonta curvipes* (at least the female), *Tropidia quadrata*, *Syritta pipiens*, have been found in California; *Polydonta* also in Colorado; *Syritta* everywhere.

Two Californian new species, intermediate between *Criorrhina* and *Brachypalpus*, I doubtfully refer to *Pocota* St. Fargeau.

Chrysochlamys has been found in Utah, but not yet in California.

An interesting discovery is that of a new *Sphecomyia* from the Sierra Nevada, the third species of the genus, or, perhaps, the second, if the European and the North American specimens belong to the same species, as may very possibly be the case. *S. vittata* has been found in the White Mountains, in the State of New York, and as far south as Virginia (probably in the mountains); recently I received it from Colorado. The European species was first discovered in Lithuania; afterward found in Finland and Norway; never in Western Europe. As far as I can judge from the figure, it seems to be the same as the North American species. The more interesting for this reason is the discovery of a decidedly distinct species, smaller, and with much shorter antennæ, found at an altitude of 7–8,000 feet in the Sierra Nevada. *Ceria* is hitherto represented in California by the single *C. tridentata* Loew.

With such scanty materials, it would be premature to draw any general conclusions about the relationship of the Californian fauna to other faunæ. As the *Syrphidæ* are among those families, the species of which are apt to have a very wide geographical distribution, the common occurrence of so many species in the Atlantic and in the Pacific States has nothing very astonishing. In the same way, many species of *Syrphidæ* are common to Europe and North America. In the occurrence of certain peculiar forms (for instance, *Eupeodes*) as well as of many species which have a wide western distribution, from California to Colorado, and are unknown in the Eastern States, the western fauna asserts its independent character.

The relationship to the European and to the Chilian fauna has hitherto shown itself only in the common occurrence of *Syrphus pyrastri*. In the occurrence of a larger number of species of *Sphærophoria*, the Californian fauna seems to resemble the European rather than the Eastern American fauna.

PARAGUS sp.—Los Angeles, Cal., in March. A single specimen.

PIPIZA sp.—Geysers, Sonoma County, May 5–7. Very like some of the *Pipizæ* described by Mr. Loew from the Atlantic States, but with darker legs than any of them.

ORTHONEURA.—Summit Station, Sierra Nevada, July 17. A single specimen.

CHILOSIA sp.—Saucelito, Cal., July 1. A single male. Exactly like *Chilosia pallipes* Loew, Centur., iv, 70 (found by me in the District of

Columbia and in the White Mountains), only larger, 8–9^{mm} long; the hind femora are altogether red, with only a slight brownish shade before the apex.

CHILOSIA sp.—Saucelito, Cal., April 2. Two males. Not unlike *Chilosia tristis* Loew from the Atlantic States.

CHILOSIA sp.—Lagunitas Creek, April 15. A single male. Eyes pubescent, and therefore different from all the species from the Atlantic States described by Mr. Loew.

As in the Atlantic States the species of the corresponding genera are yet very little known, it is not worth the while to describe Californian species here.

MELANOSTOMA TIGRINA n. sp.—Dark metallic-green; abdomen velvet-black; segments 3 and 4 each with an olive-green metallic cross-band, the first interrupted, the second connected with the olive-green hind margin of the segment; fifth segment metallic-greenish; legs black, tibiæ brownish-red. Length 8–8.5^{mm}.

Male.—Face and front metallic-greenish-black; the face with a white pollen forming transverse, irregular, dotted ripples, the intervals of which show the ground-color; the cheeks and a stripe running over the facial tubercle are bare of pollen; upper oral edge somewhat upturned and facial tubercle projecting; antennæ black, third joint and arista brown; front faintly and evenly grayish-pollinose, and with erect black pile; vertical triangle metallic-green, with black pile. Thorax metallic-bluish-green, clothed with dull grayish pile, more whitish on the pleuræ. Halteres brownish. Abdomen elongated-elliptical; first segment greenish-black; the second velvet-black, opaque, its lateral edge metallic-olive-green, with a small subtriangular expansion; third segment anteriorly with an interrupted metallic-olive-green cross-band nearly half as broad as the segment, but expanded on the sides along the whole lateral margin; the posterior half of the segment is of a velvet-black, which does not quite reach the lateral margin; the fourth segment is like the third, except that the olive-green cross-band is not interrupted, or only sub-interrupted, and connected by a longitudinal olive-green stripe, cutting through the velvet-black portion with the narrow olive-green hind margin of the segment; thus the velvet-black on this segment forms a broad interrupted cross-band, not quite reaching the lateral margin; fifth segment and hypopygium metallic-green. Femora metallic-green; knees yellowish-brown; tibiæ yellowish-brown or brownish-yellow at the base, darker toward the tip, especially the last pair; tarsi black, first joint of the middle tarsi brown; first joint of the hind tarsi incrassate. Wings subhyaline; stigma brownish-yellow.

Hab.—Saucelito, Marin County, Cal., April 2; Yosemite Valley, June 13. Two males.

The abdomen of this species is broadest at the hind margin of the second segment.

MELANOSTOMA sp.—Petaluma, California, April 28. Abdomen more linear than the preceding; face very much produced, almost conical, as in the description of *Syrphus trichopus* Thomson (Eugenies Resa, p. 502), which is evidently a *Melanostoma*. The "thorace haud viridi" of that description prevents me, however, from identifying it.

MELANOSTOMA sp.—Santa Barbara, Cal., February 10.

SYRPHUS.

I have six Californian species of this genus (in the resticted sense). They all show the greatest resemblance either to European species or to species from the Atlantic States. *S. pyrastri* is not distinguishable from the European species of that name. I have left the name of *S. lapponicus* to specimens but very little different from specimens from the Atlantic States which would pass under that name. Whether my *S. americanus* is the same as the species from the Atlantic States, the discovery of the as yet unknown female will have to prove. *S. intrudens* is remarkably like my *S. amalopis* from the White Mountains, but seems to be, nevertheless, a different species. *S. protritus* is very like the common *S. rectus* and the European *S. ribesii*, but differs in the coloring of the legs.

Syrphus affords an interesting field for the study of the limits of variation; of local, perhaps seasonal varieties. It is very desirable that entomologists should collect large numbers in both sexes, and take note of the exact date and locality of each specimen. Until the laws of variation in *Syrphus* are better known, it would be useless to multiply species upon vague and secondary characters. In the present case, as in many others, I have preferred to retain the names of European or Eastern American species whenever the Californian specimens did not show any distinctive characters which I could consider as specific.

Several of my new species I possess in the male sex only. I have, nevertheless, described them, in the hope that these descriptions, owing to the frequent coincidences with the species from the Atlantic States, would not be unwelcome from one who described the latter.

For the detailed descriptions of the eastern species, I refer to my paper, "On the North American Species of the Genus Syrphus," in the Proceedings of the Boston Society of Natural History, October 6, 1875.

Syrphus fumipennis Thomson, Eugenies Resa, 499 (California), seems to be very near my *S. americanus* and *opinator*, but does not quite agree with either. The slightly brownish tinge of the wings is not a character to be relied upon.

I. The three abdominal yellow cross-bands are interrupted (dissolved into lunate spots):

    Front very much projecting in both sexes; eyes pubescent; in the male, there is on the eyes a conspicuous area of larger facets in the middle; large species, with three pairs of sub-lunate yellowish-white abdominal spots on velvety-black ground .. .. 1. *pyrastri* Lin., ♂ ♀.

Front not unusually projecting; no conspicuous area of larger facets in the middle of the eye of the male; middle sized species, with three pairs of yellow abdominal spots, the second and third lunate:

Face with a broad brown stripe over the tubercle; eyes pubescent;* abdominal lunules deeply emarginate in the middle ....2. *intrudens* n. sp., ♂.

Face with a small brown stripe over the tubercle; eyes glabrous; abdominal lunules of nearly equal breadth,

3. *lapponicus* Zett., ♂.

II. The second and third abdominal cross-bands are not interrupted; eyes glabrous:

The second and third cross-bands do not reach the lateral margin of the abdomen:

Face with a brown stripe in the middle on the tubercle; abdominal cross-bands broad ....4. *americanus* Wied., ♂.

Face and cheeks altogether yellow; abdominal cross-bands rather narrow........................5. *opinator* n. sp., ♂ ♀.

The second and third cross bands reach the lateral margin of the abdomen; in the male, all the femora are red to the very base, coxæ and trochanters being black......6. *protritus* n. sp., ♂.

1. Syrphus pyrastri (*Syrphus pyrastri* Linné, Fauna Suecica; *Syrphus transfugus* Fabricius, Ent. Syst., iv, 306; *Syrphus affinis* Say, Journ. Acad. Phil., iii, 93, 9).

The Californian specimens do not show any difference from the European ones, which I had for comparison, except that the abdominal yellow spots are a little narrower; and even this difference does not exist in my specimen from Colorado. Macquart (Dipt. Exot., ii, 2, 83 and 88) records the same species from Chili.

It occurs everywhere in California, is not rare, and begins to appear very early. I have specimens from Santa Barbara, February 10; Santa Monica, February 18; Petaluma, April 28; San Rafael, May 29; Yosemite, June; Webber Lake, July 26; Salt Lake, Utah, August 1. I also have it from Southern Colorado (W. L. Carpenter). Say had it from Arkansas. It is very striking that a species of such wide distribution should not occur at all in the Atlantic States.

Say's synonymy is not in the least doubtful; compare especially the foot-note in Wiedemann (Auss. Zw., ii, p. 118), where he explains that *Syrphus transfugus*, to which Say compares his *S. affinis*, is *transfugus* Fabricius, a synonym of *pyrastri*, specimens of which he had sent to Say. I was wrong in connecting *S. affinis* with *S. lapponicus* in my paper on Syrphus (p. 149).

In this species, the eyes of the male have an area of large facets in the upper and middle portion; a structure which I have not observed in any

---

* The pubescence of the eyes is always easier to perceive in male than in female *Syrphi*; in the latter, a very careful examination is often required.

other *Syrphus (sensu strict.)*; the hypopygium of the male is much smaller than in *Syrphus*, entirely concealed under the fifth segment; the front remarkably convex in both sexes. These characters fully justify the formation of a separate genus, which I will call *Catabomba*, in allusion to the mode of flight of the species (from καταβομβέω, I am humming round).

2. SYRPHUS INTRUDENS n. sp.—*Male.*—Eyes pubescent; face brownish-yellow, with a broad black stripe in the middle, abbreviated before the antennæ, and narrower than the yellow portion of the face on each side of it; the black is prolonged along the oral border to the black cheeks, which have a slight greenish reflection; antennæ black, third joint sometimes slightly reddish at the base; front and vertex black, with a greenish reflection and black pile; occiput beset with a fringe of fulvous pile. Thorax dark metallic-green, clothed with fulvous pile, especially conspicuous on the pleuræ; scutellum with a shade of dull yellowish under the strong greenish-metallic luster; its pile is black; a few fulvous hairs on the sides only. Abdomen black, very little shining; on the second segment two oblong yellow spots, not reaching the lateral margin; on the third and fourth segments, a pair of deeply lunate spots, club-shaped at both ends, touching the anterior margin on one side, broadly contiguous to the lateral margin on the other; the deep excision on them has a triangular shape; fourth and fifth segments with a narrow yellow posterior margin. Halteres with a lemon-yellow knob; legs reddish; anterior femora black on their proximal half; hind femora black, except the tip; hind tibiæ with a brown ring in the middle; the other tibiæ also slightly marked with brown; tarsi brownish above. Wings distinctly tinged with brownish; stigma brownish. Length about 10ᵐᵐ.

Hab.—In the woods of the Coast Range, in the spring; Lagunitas Creek, April 15–20; also received from Mr. H. Edwards three males.

Very like *Syrphus amalopis* O. S. from the White Mountains, N. H., but the pile on the occiput is bright fulvous, not pale yellowish-white; there is more fulvous pile on the thorax; the abdominal spots are a little larger, their inner club-shaped end more clumsy, the emargination deeper; the legs are less dark-colored; the facial tubercle less prominent. Nevertheless, the resemblance is very striking. *S. amalopis* seems to be a very variable species; in the females which I have seen, the lunate abdominal spots were dissolved in two, thus forming transverse rows of four spots on segments 3 and 4; the cross-bands sometimes touch the lateral margins, sometimes not. The same variations may occur in *S. intrudens.*

3. SYRPHUS LAPPONICUS Zetterstedt, Dipt. Scand, ii, 701, 3.— In my above-quoted essay on *Syrphus* (p. 149), I have referred to this species a number of specimens from the Northern Atlantic States and the British Possessions, which agree in all respects with Mr. Zetterstedt's description, and some of which, sent by me to Dr. Loew, were

recognized by him as *S. lapponicus*. Four male specimens from California (Lagunitas Creek, Marin County, April 15; the Geysers, Sonoma
County, May 5–7; Yosemite Valley, June 5) do not differ in any essential character from the former. They are a little smaller, and the abdominal lunate spots are a little broader; the thorax is more greenish
than bluish. I have specimens from the Atlantic States, however,
which even in these particulars agree with my Californians. A specimen from British Columbia is larger, and resembles in all respects some
specimens from Maine and New York State. In Europe, the same differences occur, and there seems to exist a good deal of uncertainty about
S. arcuatus and *lapponicus*, which differ only in the degree of curvature
of the third vein. My Californian specimens have this curvature
strongly marked, but among my eastern specimens there are some
where it is very weak.

The reference to *S. affinis* Say, in my paper (l. c., 149), must be
struck out (see *ante*, in *S. pyrastri*).

4. SYRPHUS AMERICANUS (*Syrphus americanus* Wiedemann, Osten
Sacken, l. c., 145).—I provisionally refer to this species a dozen of male
specimens taken in Marin and Sonoma Counties, California, in April and
May, and in Yosemite Valley in June. They are larger than the ordinary
specimens from the Atlantic States, and measure from 9.5^{mm} to 11^{mm}.
The first yellow cross-band is more broadly interrupted; the two other
cross-bands are very variable in breadth, sometimes narrower than their
black intervals, emarginate posteriorly. I have no females, and will remark here that they have entirely yellow femora, and that the first
abdominal cross-band is, in most cases, *not* interrupted (compare l. c.,
p. 146); and, unless these characters are also to be found in the females
from California, their specific distinctness is no longer doubtful.

A specimen from Oregon (H. Edwards) is smaller, and has the yellow
abdominal spots of the first pair laterally prolonged, so as to reach the
anterior corner of the segment (instead of being entirely cut off from
the edge of the abdomen by a black margin). I am not sure about the
specific identity of this specimen.

For a detailed description of *S. americanus*, see my paper on *Syrphus*.

5. SYRPHUS OPINATOR n. sp.—*Male.*—Eyes glabrous; face, including the cheeks, altogether yellow or reddish-yellow; no brown stripe
on the facial tubercle; antennæ brown, reddish on the under side;
front above the antennæ yellow, the angle between the eyes greenish.
black, yellowish pruinose beset, with black hair; small brown marks
above the root of the antennæ. Thorax metallic-bluish-green, densely
beset with yellow pile; the broad, geminate, grayish stripe in the middle is subobsolete. Scutellum yellowish-metallic-opalescent, beset with
black pile. Abdomen black, opaque on the anterior, subopaque on the
posterior part of the segments; the two yellow spots on the second segment are prolonged anteriorly, so as to reach the lateral margin of the
segment at its anterior corner; the two other cross-bands are rather

narrow (not much broader than one-fourth of the breadth of the segment), attenuated in the middle (even subinterrupted in one of the specimens); their ends are separated from the lateral margin by a narrow black interval; posterior margin of the fourth segment with a narrow reddish border; that of the fifth still narrower. Legs reddish-yellow; proximal half (or nearly so) of the four anterior femora black; hind femora black, except at tip; hind tarsi infuscated. Stigma brownish; both costal cells distinctly tinged with brown. Length 9–11^{mm}.

Female.—I have two specimens which I refer to this species, on account of their entirely yellow face and the course of the cross-bands, which is nearly the same as in the male; but the femora are entirely reddish-yellow, coxæ and trochanters remaining black. Vertex greenish-black; front metallic-green, densely yellowish-pollinose, its lower part reddish-yellow, except two brownish marks above the root of the antennæ.

Hab.—Marin County, California (Saucelito, April 2; San Geronimo, April 20). Two males and two females.

6. SYRPHUS PROTRITUS n. sp.—*Male.*—Eyes glabrous; face yellow, with a bluish opalescence; on the cheeks a large blackish spot below the eye, and not quite reaching the oral margin (it is variable in size, sometimes very small); the lower edge of the head behind the mouth and between the lower end of the eyes is again yellow. Antennæ red, sometimes faintly brown on the upper side of the third joint; front brownish-yellow above the antennæ, black, slightly pruinose, and with black pile in the corner between the eyes; vertex black, with black pile; occiput grayish, beset with pale hairs. Thorax dark bronze-green, beset with dense yellowish pile. Scutellum yellowish, with black pile, some yellowish hairs on each side. Abdomen black, opaque, with three reddish-yellow cross-bands, the first of which is broadly interrupted; the two yellow spots thus formed are prolonged along the lateral margin to the very base of the abdomen; the second and third bands reach the lateral margin, being only a little attenuated before it; they are biconvex posteriorly, with an angular emargination in the middle; fourth segment with a yellow border posteriorly; the fifth red, with a black triangle in the middle. Legs altogether reddish, often a brown shade in the middle of the hind tibiæ and on the hind tarsi; femora red from the very base; coxæ and trochanters black. Wings subhyaline; their root tinged with brownish, the costal cell with yellowish; stigma brownish-yellow; third vein nearly straight. Length 10–12^{mm}.

Hab.—Saucelito, Marin County, Cal., April. Four males.

Is very like the common *S. rectus* of the Atlantic States; only in that species the male has black hind femora, the black spot on the cheeks is smaller, and the antennæ are much darker.

EUPEODES nov. gen.

Very like *Syrphus*, from which it can be at once distinguished in the male sex by the large development of the sixth abdominal segment and of the male hypopygium. In the female, the fifth abdominal segment is about half as long as the preceding, while in *Syrphus* proper the relation between the corresponding segments is as 1 to about 3 or 4. The scutellum in both sexes is unusually raised, exposing the metanotum more than in *Syrphus ;* in the female, it has a distinct yellow border, which is not the case in the known American species of *Syrphus.*

The sixth abdominal segment in the male is as long as the two preceding segments taken together, but narrower; it is convex, almost tubular, when seen from above, and unsymmetrical, its end pointing slightly to the right. The seventh segment on the under side of the sixth bears the opening of the anus. Beyond the anus, on the under side of the body, there are two long, linear, subparallel appendages, arcuate, bidenticulate at the end; these appendages are bent under the body when in repose, and are imbedded in a horny groove on the under side of the sixth segment, which encroaches on the fifth; when in motion, these appendages come out of the groove like a blade of a pen-knife, at an angle to the axis of the abdomen; in length, they are nearly equal to the whole sixth segment.

The name has reference to the structure of the hypopygium.

EUPEODES VOLUCRIS n. sp.—*Male.*—Eyes bare. Face whitish-yellow, with black cheeks and a brown stripe over the facial tubercle; front whitish-yellow, with some black pile; antennæ dark brown; vertex black. Thorax dark metallic-green, sometimes slightly bluish, with very pale yellowish pile; scutellum yellowish, more or less metallescent, with pale yellow pile; abdomen black, opaque; the first segment, the lateral and posterior margins of all the segments, shining; the fifth altogether shining; on the second segment two yellow oblong spots, well separated from the lateral margin; on each of the two following segments, a pair of larger, oblong, yellow spots; those on segment 3 very slightly lunate; the posterior margins of the fourth and fifth segments with narrow yellow margins. The sixth segment is black, shining, sparsely beset with whitish pile; its shape has been described in the generic character. Legs reddish; base of femora black; hind femora black, except the tip; hind tarsi more or less brown on the upper side. Wings hyaline; stigma yellowish-brown; anterior costal cell hyaline, the posterior tinged with yellowish.

Female.—Front and vertex black; across the black a faint subinterrupted arcuate stripe of pollen, leaving a triangular glabrous black space below; lower part of the front yellow, except a dark brown crescent-shaped spot above the root of the antennæ; a narrow yellow space between this spot and the black above. Scutellum distinctly yellow along the edge; the black at the root of the femora is a little less extensive here. Length very variable, from 10^{mm} down to 7^{mm}.

Hab.—California, Nevada, Utah, Colorado, common. I found it commonly in Los Angeles in February; in Marin County in April; in Yosemite in June; in Utah in August. I also have specimens from Denver, Clear Creek, etc., Colo., August (P. R. Uhler); Spanish Peaks, Colo., June 15 (W. L. Carpenter). More than two dozen specimens of both sexes.

MESOGRAPTA GEMINATA (Say), Journ. Acad. Phil., iii, 92, 7 (*Scæva*).— Occurs both in the Atlantic and Pacific States (San Rafael, Cal., April, May ; Yosemite, June).

MESOGRAPTA MARGINATA (Say), Journ. Acad. Phil., iii, 92, 6, (*Scæva*).— Common on both coasts (Los Angeles, Cal., in February ; Webber Lake, Sierra Nevada, in July); also in Denver, Col. (Uhler). Is not *Syrphus limbiventris* Thomson (Eugenies Resa, 495) simply a variety of this species ?

SPHÆROPHORIA SULPHURIPES (Thomson), Eugenies Resa, 500 (*Syrphus*).—Specimens (♀) from San Rafael, May 29, and Yosemite, June 14, agree with Mr. Thomson's description. The cross-band on the fourth segment is sometimes interrupted. Whether the male specimen described by Mr. Thomson belongs here seems doubtful. I have males with entirely yellow coxæ, like those of the female ; the cross-bands or segments 2 and 3 are not interrupted, and reach the lateral margin ; segments 4–6 are reddish, with brownish marks. In other males, the hind coxæ are dark, but with a yellow spot behind; the cross-bands are laterally interrupted before reaching the margin. I also have specimens with a brown stripe over the face, dark femora, and hypopygium.

California seems to be rich in species of this group, richer than the Atlantic States; and in this it again resembles Europe. In Europe, the definition of the species of *Sphærophoria* is, as yet, an unsolved problem ; they seem to be very variable in their coloring, and it would not be safe to multiply descriptions of Californian species based on coloring only. It seems doubtful to me whether the *Syrphus infuscatus* Thomson is not the same species as his *sulphuripes*, and I am not at all sure whether the latter is not identical with the common *Sphærophoria cylindrica* of the Atlantic States.

Among the several species of this genus which I have before me, I will describe only one, which has very marked characters to distinguish it.

SPHÆROPHORIA MICRURA n. sp.—*Male.*—Face of a somewhat livid yellow, with a brown stripe in the middle; front above the antennæ, with a large semicircular greenish-metallic spot; the interval between this spot and the eyes is yellow; antennæ brown, third joint reddish at the base and on the under side; cheeks metallic blackish-green, but oral margin yellow. Thorax dark metallic-green, with the usual antealar humeral yellow stripes; scutellum yellow, with black pile; pleuræ dark metallic, somewhat bluish; abdomen black; first segment with a very narrow basal yellow margin; segments 2–4 each with a straight yellow cross-band, reaching the lateral margin, and framed in anteriorly and posteriorly in velvet-black, opaque cross-bands; the hind margins of the segments are shining bluish-black ; the cross-band on segment 2 is narrower than the two others, and sometimes narrowly interrupted in the middle; segment 5 has yellow sides and two yellow streaks in the middle; the hypopygium is black, and unusually small for a *Sphærophoria*. Legs dark brown or black; the ends of the middle and front femora to a greater or less extent brownish-yellow. Wings with a distinct brown tinge. Length 7.5–9^{mm}.

Hab.—California (the Geysers, Sonoma County, May 5–7; San Rafael, May 29; Brooklyn, near San Francisco, July 11). Seven males.

Easily distinguished by the very small, black hypopygium, the color of face and front, the dark legs, etc. The dark metallic color of the cheeks sometime extends along the interval between the occiput and the posterior oral margin; sometimes there is in that interval a more or less extensive yellow spot.

ALLOGRAPTA FRACTA n. sp.—*Male.*—Face, including the frontal triangle, pale yellow, slightly opalescent; a bluish-black stripe extends from the oral edge to the antennæ, forming a semicircle above them; antennæ reddish, third segment brown along the upper edge; vertex black. Thorax bright metallic green, a pale yellow stripe on each side between the humerus and the root of the wings; antescutellar callosity yellowish; scutellum of a saturate yellow, the extreme corners dark; halteres with yellow knobs. First abdominal segment metallic greenish-black, its extreme anterior margin only yellow; the rest of the abdomen black, opaque; an interrupted yellow cross-band on the second segment, equal to about one-third the segment in breadth; a somewhat broader, slightly arched, and not interrupted yellow cross-band on the third segment; on the fourth, two narrow, parallel, longitudinal lines in the middle, and an obliquely placed, large, oval spot on each side of them, yellow; the narrow fifth segment shows a yellow picture, somewhat resembling that of the fourth segment. Legs yellow; tips of tarsi brownish; hind femora with a brown ring before the tip; hind tibiæ with two such rings, one before the middle, the other before the tip; hind tarsi brown, except the under side of the first joint. Wings hyaline; stigma brownish-yellow. Length 7^{mm}.

Hab.—Santa Monica, Cal., February 20, 1876. A single male.

Observation.—I perceive, even in the dry specimen, the difference between the larger facets of the upper half of the eye, and the smaller ones of the lower half, a character which I have pointed out as distinctive of the new genus *Allograpta* (see Buff. Bull. N. H., iii, 49). The coloring of the abdomen of this species is similar in character to that of the typical *A. obliqua* Say.

SPHEGINA.—A single female, from Lagunitas Creek, Marin County, California, April 15, black, with a red abdomen and red legs, seems to differ from *S. infuscata* Loew from Sitka (Centur., iii, 23).

BACCHA LEMUR n. sp.—Wings hyaline, with an incomplete brown cross-band between the stigma and the fourth posterior cell; abdomen with two red cross-bands. Length 10–11^{mm}.

Front and vertex metallic greenish-black, the former (in the ♀)

10 H B

whitish-pruinose along the eyes; face whitish-pruinose, its ground-color variable, dark metallic-green, with more or less brownish-yellow on the sides and on the facial tubercle, or entirely yellowish; antennæ brown or reddish-brown, inserted on brownish-yellow ground.   Thorax metallic greenish-black, with vestiges of whitish-pruinose stripes anteriorly; pleuræ whitish-pruinose, with white pile; scutellum translucent yellowish-brown, with a metallic reflection; halteres with yellow knobs.   Abdomen black, shining, with bronze and bluish reflections; a broad blood-red cross-band slightly emarginate in the middle posteriorly, at the base of the third and fourth segments; these cross-bands are slightly pruinose; the sides of the abdomen beset with white pile.   Legs pale yellow; hind femora and tibiæ usually each with a brownish ring; sometimes the legs are more brownish, especially on the femora.   Wings hyaline; the root before the humeral cross-vein and the extreme base of the second basal cell is infuscated; costal cell hyaline, but the interval between auxiliary and first veins is pale brownish; stigma dark brown, the corner between the costa and the end of the first vein yellowish; a brown, incomplete cross-band between the first and fifth veins; it coalesces with the brown stigma, leaves hyaline the extreme proximal end of the submarginal cell, covers the small cross-vein and the cross-vein at the base of the discal cell, but reaches only very little beyond either, and ends at the fifth vein, filling up the proximal end of the fourth posterior cell.

Hab.—Santa Monica, Cal., February 18; Summit Station, Sierra Nevada, July 17; Fort Bridger, Wyo., August 4; Morino Valley, New Mexico, July 1, W. L. Carpenter.   One male and three females.

BACCHA ANGUSTA n. sp.—Wings hyaline at the base, slightly shaded with brownish-gray beyond the cross-veins; abdomen with two yellow cross-bands.   Length 7-8^{mm}.

Male.—Face and front metallic-green; antennæ brownish (the head is somewhat injured in my specimen.)   Thorax and scutellum bronze-color; halteres brownish-yellow; the tip of the knob brownish.   Abdomen: two first segments bronze-color, the second long and very slender, the remainder metallic-brown, with brownish-yellow cross-bands at the base of the third and fourth segments; the cross-band on the fourth segment occupies about one-third of its breadth; that on the third is a little narrower.   Wings hyaline from the root to the central cross-veins, slightly tinged with grayish beyond; a brown cloud between the tip of the auxiliary vein and the first vein; beyond this cloud, the space between the costa and the first vein is brownish-yellow.   Legs yellowish, more or less tinged with brown in the hind pair.

Hab.—Lagunitas Creek, Marin County, California, April 15.   A single male.

VOLUCELLA MEXICANA Macquart.—Besides Mexico and Texas, this species occurs in Southern California.   I have received specimens from the island of Santa Rosa (S. Cal.), through the kindness of Mr. H. Edwards.

VOLUCELLA MARGINATA Say, Journ. Ac. Phil., vi, 166.—Mexico (Say); Waco, Texas (Belfrage); a specimen was kindly given by Mr. E. Burgess.

VOLUCELLA AVIDA n. sp.—Face with a narrow brown stripe; cheeks shining, black; abdomen honey-yellow; hind margins of segments and longitudinal narrow dorsal stripe on segments 2–4 black; wings hyaline; cross-veins and stigma clouded with dark brown. Length 11–12^{mm}.

Antennæ, light-brown; arista of the same color, plumose; third antennal joint about two thirds of the length of the arista, its basal half a little expanded; face pale whitish-yellow, a narrow black stripe runs from the mouth upward, becoming paler and finally obsolete before reaching the antennæ; cheeks black, shining; frontal triangle of the male pale whitish-yellow, beset with black pile; profile straight, with a slight depression under the antennae; eyes densely pubescent. Thorax greenish-black; on each side, between the humerus and the scutellum, a rather broad, dull, honey-yellow stripe; in front of the scutellum, a yellowish, rather obscure parallelogram, emarginate anteriorly; sides of the dorsum and pleuræ beset with yellowish white hairs. Scutellum yellowish, subtranslucent, beset with black hair along the edge. Abdomen pale honey-yellow; first segment black; the second and third segments posteriorly with a narrow black margin, expanded on each side along the lateral margin, and prolonged in the middle in the shape of a lon-gitudinal black stripe toward the anterior margin; on segment 2, this stripe is broadly expanded, so as to coalesce with the black of the first segment; on segment 3, on the contrary, it is tapering anteriorly; segment 4 with a black cross-band a little beyond the middle, with a subtriangular expansion in the middle, reaching toward the anterior margin; hypopygium, black; the abdomen is clothed with short black hairs on its black portions, and with longer yellowish-white pile in the yellow regions, especially on the sides and around the black triangle on the second segment. Femora black; knees and anterior half of the tibiæ brownish-yellow; tarsi black. Wings, hyaline; latter half of the subcostal cell and the interval between the auxiliary and first longitudinal vein as far as the stigma brownish; stigma dark brown; central cross-veins and small cross-vein with well defined, although small, brown clouds; the second vein ends in the first, some little distance before the tip of the latter.

Hab.—California (G. R. Crotch). A single male.

VOLUCELLA SATUR n. sp.—Face altogether yellow; cheeks yellow, except a black stripe from the lower corner of the eye to the anterior oral edge; abdomen honey-yellow; first segment black, hind margin of segments 2 and 3 with a narrow black border, that of the second segment connected with the black first segment by a broad black stripe expanding anteriorly; wings hyaline, cross-veins and stigma clouded with pale brown. Length 9–10ᵐᵐ.

Antennæ light brown; arista reddish, plumose; third antennal joint nearly as long as the arista, linear. Face yellow, a black stripe runs obliquely from the lower corner of the eye to the anterior oral margin; behind it the cheeks are yellow; profile of the face straight; the depression under the antennæ is hardly perceptible; the frontal triangle of the male is yellow, beset with black hair, th e vertex black; in the female, the front has a greenish tinge, as if underlying the yellow; a slender yellow line runs from the antennæ toward the yellow vertex; the ocelli are placed on a cordiform black spot. Eyes densely pubescent. Thorax blackish-green; on each side, between humerus and scutellum, a rather broad, dull, honey-yellow stripe, with a short black streak in the middle; in front of the scutellum, a yellowish, rather obscure parallelogram, emarginate anteriorly. Pleuræ with a large yellow spot under the humeri; they are beset with yellow pile. Scutellum yellow, with black pile on the edge. Halteres with yellow knobs. Abdomen honeyyellowish; first segment black; second and third with a narrow, parallel, black hind border; the black border of the second segment is connected with the black of the first segment by a black longitudinal stripe, which is narrow in the female, broad and triangularly expanded anteriorly in the male; fourth segment with a broader black hind border; the fifth black. Femora black; knees and two-thirds of the tibiæ brownish-yellow; the last third black, or, on the intermediate pair, brownish; tarsi reddish at base, brownish or black at tip. Wings grayish-hyaline; stigma yellowish, with a small, pale brown cloud; cross-veins at the base of the first and last posterior cells and of the discal cell and the origin of the third vein with small brown clouds; still smaller, almost imperceptible clouds at the tip of the second vein, near its junction with the first, and on the curvature of the vein closing the first posterior cell; this curvature is much less strong here than in *V. fasciata;* the second vein ends in the first close by the tip of the latter.

Hab.—Colorado Plains (W. L. Carpenter). I took a specimen in the railway-carriage, between Wahsatch and Evanston, Utah, at an altitude of 6,800 feet, August 3. Two males and one female.

VOLUCELLA FASCIATA Macquart, Dipt. Exot., ii, 2, 22.—Occurs in Texas; also in Manitou, Colo.

TEMNOCERA SETIGERA n. sp.—Proboscis nearly twice as long as the head, pointed at the end; snout projecting in the shape of a cone; scutellum with fourteen black bristles along the edge; abdomen brownish-yellow, with a black spot at the tip, embracing segment 5 and a part of segment 4. Length 14ᵐᵐ.

Female.—Face and front honey-yellowish, clothed with black pile, which is very short on the face and longer on the front; the face is excavated below the antennæ, its lower part projecting in the shape of a cone, the tip of which is bifid and slightly infuscated. Antennæ: first two joints yellowish-brown; third joint light brown, excised above, so that its latter portion is much narrower; arista feathery, black; proboscis 7-8^{mm} long, black, pointed. Thorax densely clothed with a yellowish recumbent pubescence, and, mixed with it, short, black, erect pile; they almost conceal the dark greenish ground-color, as well as the obscurely visible yellowish lateral stripes and large yellowish spot in front of the scutellum; on the sides of the thorax, several stiff, black bristles; a pair of such bristles, but smaller, a little in front of the scutellum; pectus blackish. Scutellum somewhat inflated, honey-yellow, beset with a mixed black and yellow pubescence; along the edge fourteen stiff black bristles. Abdomen brownish-yellow; second and third segments with broad blackish parallel borders posteriorly, formed by short and very dense black hairs; the cross-bands thus produced are very distinct when viewed obliquely, although almost invisible from above; that on the second segment occupies more than one-third, that on the third more than one-half of the length of the segment; segment 4 shows posteriorily a semicircular, black, shining spot, occupying the whole posterior margin and reaching beyond the middle of the segment anteriorly; segment 5 is black. Femora black; knees and anterior half of the tibiæ brownish-yellow; the remainder of the tibiæ as well as the tarsi are darker. Wings grayish-hyaline; cross-veins and tip of second vein with small brown clouds; stigma brown.

Hab.—Vermejo River, New Mexico, June 25 (W. L. Carpenter).

I do not hesitate to describe this well-marked species, although I have only a single, not very well preserved female. On account of the bristles on the scutellum and the shape of the third antennal joint, I place it in the genus *Temnocera*, although I do not think that this genus is defined in a very satisfactory manner.

TEMNOCERA MEGACEPHALA Loew, Centur., iv, 57.—California. I do not know this species.

ARCTOPHILA FLAGRANS Osten Sacken, Bulletin Buffalo Soc. N. H., iii, 1875, 69.—*Male.*—Face wax-yellow. Cheeks black. Antennæ: basal joints brownish; third joint reddish, the plumose arista black. Thoracic dorsum densely clothed with yellowish hair, through which, however, the metallic brownish-coppery ground-color is apparent; pleuræ black in the middle, with a stripe formed by yellow pile. Abdomen with long yellow pile at the base and on the sides, with reddish hair in the middle and at the tip; between the hairs, the black metallescent ground-color is apparent. Legs black; front tibiæ beset on the inside with short golden-yellow hairs; three basal joints of the four posterior tarsi brownish-red. Wings with a slight grayish tinge; a brown spot limited by the fourth longitudinal vein, the costa, the small cross-vein, and the origin of the third vein; the latter vein is more deeply sinuate than in *A. bombiformis.* Length 13^{mm}.

Hab.—Colorado Mountains (Lieut. W. L. Carpenter). A single male.

ERISTALIS HIRTUS (*Eristalis hirtus* Loew, Centur., vi, 66; *Eristalis temporalis* Thomson, Eug. Resa, 490.)—Common in California as well as in the Rocky Mountains (environs of San Francisco, May, June; Yosemite, June; Lake Tahoe, July 19; Webber Lake, July 27; Georgetown, Colo., August 12). I have also specimens from South Park and Twin Lakes, Colorado, by W. L. Carpenter.

The specimens vary in size from 5^{mm} to 8^{mm}. As a rule, those from high altitudes are smaller and darker in color. To Mr. Loew's description of the male, otherwise remarkably accurate and complete, should be added that the frontal triangle above the antennæ is rather convex, and bears a conspicuous tuft of yellowish pile; on the anterior part of the fourth abdominal segment, in the middle, there is a velvet-black streak, similar to a corresponding streak on the preceding segment; and, instead of "in segmentorum *tertii* et *quarti* partibus nigrovelutinis", read "secundi et tertii".

The female has a remarkably broad and convex front, a very characteristic mark of the species; it bears a dense crop of yellowish hair. The black vertex has some black pile in the middle. The yellow triangles on the second abdominal segment are usually smaller than in the male; in many specimens, they are subobsolete, brownish; often the yellow disappears entirely, leaving only two shining black triangles on velvety-black ground. On the wings, there is, in some specimens, a brownish shadow in the middle, immediately beyond the central cross-veins. The specimens with the obsolete and subobsolete yellow abdominal triangles seem to come principally from the higher altitudes.

I have seven males and twenty females.

ERISTALIS STIPATOR n. sp.—Eyes pubescent, the yellow arista bare; second abdominal segment with a yellow triangle on each side, framed in posteriorly by a velvety-black cross-band, interrupted (or subinterrupted) in the middle; narrow posterior margins of segments 2–4 yellowish-white, beset with a rather conspicuous fringe of pale golden-yellow, comparatively long hairs, this fringe being broadest on the fourth segment. Length 9^{mm} to 13^{mm}, sometimes larger.

Male.—Face yellowish-white, densely clothed with hairs of the same color; the black stripe in the middle is rather broad; cheeks black, shining; antennæ black, third joint dark brown; arista reddish-yellow, glabrous; eyes pubescent, the suture between them rather short (about half as long as the interval between apex of the frontal triangle and the root of the antennæ), the apex of the vertical triangle being considerably prolonged in front of the antennæ. Thorax greenish-black, unicolorous, shining, beset with yellowish pile, which is denser on the pleuræ. Scutellum reddish-brown. Second abdominal segment with a yellow triangle of the usual shape on each side; a velvet-black cross-band on the anterior margin, another one along the posterior side of the yellow triangles; the latter is interrupted (or subinterrupted) in the middle, oblique on each side, and interrupted before reaching the lateral margin; a smooth bluish-black space is inclosed between the two cross bands and the triangles; a narrow, shining, triangular space between the last cross-band and yellowish-white posterior margin of the segment, which bears a fringe of pale golden-yellow hairs; the third and fourth segments have the same pale yellowish posterior margin and the golden fringe upon it; on the fourth, however, the fringe is broader, and takes in the whole posterior half of the segment; on the posterior half of the third segment, there is on each side an elongated velvet-black spot; the anterior margin of this segment has a narrow, pale border, as if prolonging the hind margin of the preceding segment; hypopygium black. Legs black; tip of the femora and anterior half of the tibiæ yellowish-white; on the middle pair, three quarters of the tibiæ and the base of the tarsi are of a pale color. Wings hyaline; stigma small, brown.

Female.—Front broad and rather convex, grayish-pollinose, beset with a dense grayish-white down; vertex a little darker; no velvet-black spots on the third segment; lateral abdominal triangles often brownish-yellow; sometimes a reddish-brown shade in the middle of the wing; for the rest, like the male.

Hab.—Manitou Park, Colorado (P. R. Uhler); Morino Valley, New Mexico, July 1 (W. L. Carpenter); Denver, July 10 (A. S. Packard); California (G. R. Crotch). Four males and eight females.

This species is very variable in size; the four males and four females from Manitou being only 9–10ᵐᵐ long. My only specimen from California has the thoracic pile more reddish, that on the face more yellowish.

ERISTALIS sp.—California (H. Edwards). Very like *E. bastardi* of the Atlantic States, but different in the more metallescent surface of the abdomen and the presence of two grayish thoracic stripes in the female, abbreviated posteriorly. Some specimens from Vancouver Island seem also to belong here. As the species seems to be variable, I do not attempt to describe it with the insufficient material which I have on hand.

ERISTALIS ANDROCLUS Walker, List, etc., iii, 612.—The species which, rightly or wrongly, Mr. Loew and myself have identified with Mr. Walker's description, has a very wide distribution. It occurs in Canada, in the White Mountains, in Western New York (Cayuga Lake). I found several specimens near Ogden, Utah, August 2, 1876. Specimens from Yukon River, Alaska, have the arista dark and the velvety spots on the abdomen somewhat different.

HELOPHILUS LATIFRONS Loew, Centur., iv, 73.—My Californian specimens agree with Dr. Loew's original specimens, and also with his description, except the words "hypopygium maris plerumque flavum". In all my specimens, including Dr. Loew's originals, the black ground-color of the hypopygium is concealed under a thick yellowish-gray pollen, and is beset with yellow pile.

H. latifrons (male) differs from *H. similis* (male) of the Atlantic States in the greater breadth of the front (it is at least by one-half broader), the darker antennæ, the broader yellow cross-bands on the abdomen, leaving a narrower black posterior margin of the segments.

I have five males from Petaluma, Sonoma County, Cal., April 28. Mr. Loew described specimens from Nebraska (F. V. Hayden). I have also seen some from the Red River of the North (R. Kennicott).

HELOPHILUS POLYGRAMMUS Loew, Centur., x, 55.—The author describes the female. In the male, the front is but very little narrower than in the female; the color of the abdomen is lighter yellowish-brown on the sides, especially on the second and third segments.

Hab.—Webber Lake, Sierra County, California, July 27. A male and a female. Oregon (H. Edwards).

MALLOTA POSTICATA Fabricius, Syst. Antl., 237, 21 (*Eristalis*).—I took a male specimen near San Rafael, Cal., May 29, which resembles this species very much. Unfortunately, I have only a single damaged male from the State of New York for comparison. In the Californian specimen, the eyes do not come in contact, as there is a very narrow frontal interval between them. There is a brown cloud in the middle of the wing, especially on the central cross-veins, which does not exist in my eastern specimen. These differences render the specific identity uncertain.

Macquart is wrong when he calls the eyes of the male pubescent. My statement (in the Bull. Buffalo Soc. N. H., Dec., 1875, 64), that the male has a projecting spur in the middle of the hind tibiæ, is likewise erroneous; it was based upon a specimen in which the pubescence of the hind tibiæ was clotted, so as to produce the appearance of a spur.

POLYDONTA CURVIPES (Wiedemann; synonym in the male sex with *P. bicolor* Macq.; in the female with *Helophilus albiceps* Macquart, Dipt. Exot., 1er suppl., 132, 9).—The male of this species is most remarkably different from the female. A female in the Mus. Comp. Zoöl., Cambridge, Mass., from San Francisco, Cal. (W. Holden), resembles the eastern specimens; only the face is more whitish than yellowish, and the vertex a little less thickly yellowish-pruinose. I also have received specimens from Northern New Mexico (W. L. Carpenter). I would not pronounce on the identity of these western specimens before seeing the males.

TROPIDIA QUADRATA (Say), a male from Marin County, California (H. Edwards), does not differ from specimens from the Atlantic States.

POCOTA ALOPEX n. sp.—Black; thoracic dorsum with dense yellowish-rufous pile; pleuræ black; wings tinged with reddish-brown anteriorly, subhyaline posteriorly. Length 10–11^{mm}.

Female.—Antennæ brown, first joint black; arista rufous; head black, shining; front rather broad, beset with yellow pile. Thoracic dorsum beset with dense yellowish-rufous pile, which nearly conceals the shining black, submetallic ground-color; pleuræ black, with black pile; scutellum black, with a purplish reflection and long black pile along the edge; halteres brownish. Abdomen black, shining, beset with black pile and some scattered pale yellow pile toward the tip.

Legs black; knees and base of tibiæ of the two anterior pairs pale brownish; hind femora somewhat incrassate and beset with a tuft of yellow hairs above; hind tibiæ rather stout. Wings tinged with reddish-brown anteriorly, especially along the veins, the inside of the cells being paler; posterior portion subhyaline, slightly brownish along the veins.

Hab.—Marin County, California (H. Edwards). A single female.

Observation.—If a face prolonged downward, and provided with a tubercle in the middle, is to be considered as characteristics of *Criorrhina*, the present species and the following do not belong in that genus. In both of these species, the face forms a short snout, prolonged anteriorly rather than downward, somewhat obtusely keel-shaped above, and deeply emarginate at the tip. There is no tubercle on the face, which is in the profile gently concave between the antennæ and the oral edge. The hind femora are stouter than in *Criorrhina*, especially a short distance before the tip. As these species have no spines on the under side of the hind femora, and as the palpi are rather long and narrow, they cannot be placed in the genus *Brachypalpus*. The great looseness with which most of the genera of *Syrphidæ* are defined makes me very little inclined to increase their number without absolute necessity. I prefer, therefore, to place these species provisionally in the genus *Pocota*, adopted by St. Fargeau and resuscitated by Schiner, the face of which is without tubercle, although, judging from the description, it has a somewhat different structure. I will observe here, at the same time, that Schiner calls the genus *Plocota*, while I find *Pocota* in the Encyclopédie Méthodique, probably from πόκος, sheep-wool; ποκόω, to cover with wool.

POCOTA CYANELLA n. sp.—Thorax greenish-black, beset with long, grayish pile above, and yellowish-white pile on the sides; abdomen dark bluish-metallic; in the male with a black, opaque second segment, and a black, opaque cross-band on the third; legs black. Length 9–10^{mm}.

Face black, shining; snout projecting, keel-shaped above; on each side of the snout, a broad stripe of grayish pollen somewhat conceals the black ground-color; front and vertex black, but little shining in the female, and densely clothed with yellowish-white pile, some of which descends along the sides of the face, below the antennæ; the occiput and the posterior and inferior orbits of the eyes are beset with pile of the same color. Antennæ brown; first joint paler; arista reddish. Thorax metallic greenish-black, densely clothed with pile, which is of a dull-grayish on the dorsum and more yellowish-white on the pleuræ. Halteres brownish. Abdomen dark metallic-blue, beset, especially on the sides, with whitish pile; in the male, the second segment is black, opaque, except two triangles on each side, which are metallic-blue; the third segment has an arcuated, black, opaque cross-band, somewhat interrupted in the middle. Legs black; tibiæ brown; knees of the two first pairs yellowish-brown; the legs are beset with whitish pile, which is longer on the femora, short on the tibiæ; hind femora somewhat incrassated; hind tibiæ rather stout and somewhat curved. Wings sub-hyaline, grayish.

Hab.—Santa Barbara, Cal. Two males and one female. I reared these specimens from pupæ which I had found under the bark of an evergreen oak (*Q. agrifolia*), in February.

SYRITTA PIPIENS Lin. is common in California, Nevada, and Colorado.

CHRYSOCHLAMYS.

I possess three North American species of this genus, all of which seem to be different from the only North American species hitherto described, *C. buccata* Loew. The four species may be tabulated as follows :—

Arista black. ....4. *crœsus*.
Arista reddish :
    Legs entirely reddish-yellow ....2. *dives*.
    Anterior femora at base and tips of all the tarsi black...1. *buccata*.
    All the femora brown; tibiæ likewise infuscated........3. *nigripes*.

As the resemblance in the coloring of all these species is very great, I will first insert Mr. Loew's description, and then describe the other species by merely stating the differences.

1. CHRYSOCHLAMYS BUCCATA Loew, Centur., iv, 72.—*Female.*
Translation.—"Bronze-colored; scutellum testaceous; front black, with a cross-band of ochraceous pollen in the middle; antennæ black above, rufous below; arista rufous; cheeks with a black stripe. Length $3\frac{7}{12}$ Rhenish lines (a little less than 8^{mm}); length of wing $3\frac{5}{12}$ lines. Blackish bronze-colored, greenish ('aeneo-nigra, subvirens'), shining. Front deep black, shining, with a short, black pubescence; in the middle a rather broad cross-band of ochraceous pollen. Antennæ rather large; first joint deep black, the following joints black above, rufous below; arista glabrous, rufous. Face testaceous-yellowish, very concave and ochraceo-pollinose above, swollen ('buccata') below, with a large obtuse tubercle, which is somewhat brownish; cheeks separated from the face by a small black stripe. Thoracic dorsum beset with short, lutescent pile; lateral margins and two longitudinal stripes of even breadth, cinereo-pollinose. Scutellum testaceous, beset with short lutescent pile; bristles along the edge black; lateral corners blackish. Abdomen shining, with short lutescent pile; hind margins of segments 1 and 2 deep black, opaque. Legs ochraceous, the proximal half of the four anterior femora and the tips of all the tarsi black; front tibiæ, except their base and tip, and the base of the hind femora, slightly subinfuscated. Wings somewhat cinereous-hyaline, lutescent near the base; costal cell and stigma luteous, the base of the third vein and the cross-veins on the middle of the wing clouded with black."

Hab.—Virginia.

2. CHRYSOCHLAMYS DIVES n. sp.—*Male and female.*—Very like the

preceding (as far as I can judge from the description), but the female is larger, face altogether beset with ochraceous pollen, except on the tubercle and on the cheeks; the brownish mark on the tubercle is V-shaped; in the male, the front is black, shining above the antennæ only, pollinose along the eyes. Abdomen of a pure bronze-color, densely beset with golden-yellow pile in the female; somewhat darker in the male; the velvet-black hind margins of segments 2 and 3 are subinterrupted in the female, and somewhat broader in the male. Legs of the female of a saturate reddish-yellow, the penultimate tarsal joint slightly infuscated; in the male, the tips of the tarsi are infuscated and the four anterior femora have a brown spot on the front side. Wings yellowish-hyaline on the antero-proximal portion, grayish-hyaline along the posterior margin; costal cell yellow; stigma saturate-yellow; a brown cloud, in the shape of a short cross-band, between the root of the third vein and the cross-vein at the base of the last posterior cell; small cross-vein likewise clouded with brown (the coloring of the wing is more intense in the female than in the male). Seems to be very variable in size; one of the males is about 10^{mm} long, the other 8^{mm}; the female nearly 12^{mm}.

Hab.—Kentucky (F. G. Sanborn). Two males and one female.

3. CHRYSOCHLAMYS NIGRIPES n. sp.—*Male and female.*—General coloring much duller than in *Chr. dives*, metallic blackish-green; pollen on the face and front dull yellowish; frontal pollinose cross-band (♀) much narrower, and hence the black, shining space above the antennæ larger. The prevailing pubescence on thorax and scutellum is black; black, opaque hind margins of the segments 2 and 3 in the male only, *not in the female.* Femora brown, except the tip; tibiæ brownish-yellow, more or less infuscated before the tip, especially the front pair; tarsi brownish-red at base, brown at tip. Wings grayish-hyaline, feebly tinged with brownish-yellow at the root and on the stigma; costal cell subhyaline in the female; brownish clouds on cross-veins very weak. Length about 9^{mm}.

Hab.—Massachusetts (F. G. Sanborn). A male and a female on the same pin.

4. CHRYSOCHLAMYS CRŒSUS.—*Male.*—Very like *C. dives*, but differs in the arista being black and the hind margins of the abdominal segments 2 and 3 without velvet-black hind borders. The brown spot on the facial tubercle is of an indefinite outline, not V-shaped, as in the two preceding species. Antennæ reddish-brown, very little darker along the upper edge. Abdomen uniformly of a bright bronze-green, thickly beset with golden-yellow pile. Legs altogether of a saturate reddish-yellow. Length 10-11^{mm}, but much broader than the male of *C. dives* of the same size.

Hab.—Near Salt Lake City, Utah (Mr. Barfoot).

SPHECOMYIA BREVICORNIS n. sp.—*Male.*—Antennæ black, about half as long as the eye from its upper to its lower corner; they are inserted on a conical, black projection of the front; joints nearly of the same length, the first cylindrical, the second subtriangular, the third rounded, somewhat brownish; arista yellowish. Face and front golden-yellow, the former with a black stripe reaching from the antennæ to the mouth; cheeks black; vertex black; posterior orbits golden-yellow. Thorax black; humeri, two dorsal lines, interrupted in the middle and not reaching the scutellum, a large spot on the pleuræ and a smaller one under it yellow; scutellum yellow, its posterior edge black, beset with brownish pile; halteres with yellowish knobs. Abdomen yellow; first segment black at base; the second segment has two narrow black cross-bands, the one at the base, the other about the middle; the second does not reach the lateral margins; they are connected in the middle by a black line; the third segment has a narrow black border anteriorly, a small, black, diamond-shaped spot in the middle, and two black streaks on each side between this spot and the lateral margin; the black anterior margin of the fourth segment is entirely concealed under the preceding segment, but a diamond-shaped black spot in the middle and black streaks on the sides are similar to those of the preceding segment; hypopygium yellow. Femora black, except the tip, which is yellowish; the hind femora have the latter half brownish-yellow; tibiæ and tarsi brownish-yellow; the two last joints of the tarsi black; the end of the third joint brown. Wings tinged with brownish, somewhat yellowish at the base and along the anterior margin; a brownish cloud on the cross-veins. Length 11–12^{mm}.

Hab.—Webber Lake, Sierra County, California, July 27. A single male.

This species is very like the well known *Sphecomyia vittata*, but is smaller, has the two first joints of the antennæ much shorter, and a Somewhat different picture of the third and fourth abdominal segments; the femora are darker. In other respects, the resemblance is great. It is not improbable that the female has a somewhat different abdominal picture.

SPHECOMYIA VITTATA has been brought from Southern Colorado by Lieut. W. L. Carpenter.

CERIA TRIDENS Loew, Centur., x, 57.—A male from Sierra Nevada, California (H. Edwards), agrees with the description, except that the hind tarsi are yellowish at the base.

Family MYOPIDÆ.

My Californian collection contains species of the genera *Conops*, *Myopa*, and *Zodion*.

MUSCIDÆ (in the widest sense).

In this large division, I will confine myself for the present to the publication of a few species belonging to the *Ortalidæ* and *Trypetidæ*, the two families so thoroughly worked up by Dr. Loew in the third volume of the Monographs of the North American Diptera. To the small number of the western species which I am able to describe now, I add a few interesting new species, recently discovered in the Atlantic States·

A very striking *Dejeania*, very common in the Rocky Mountains of Colorado, deserves to be described at once, in order to draw the attention of collectors to its habits. It is very remarkable that *Dejeania*, a South American and Mexican genus, should occur so commonly at high altitudes in the Rocky Mountains among alpine forms, and it would be worth the while to investigate on what insect (probably Lepidopterous) it preys as a parasite.

DEJEANIA VEXATRIX n. sp., ♂ ♀.—Head and thorax brownish-yellow; abdomen bright ferruginous red, with a reddish-yellow pubescence and with black spines; legs red; wings pale brownish. Length 12–13^{mm} (exclusive of the length of the bristles, antennæ, etc.).

Face and cheeks pale yellow; cheeks with some long and soft fulvous pile; front brownish-yellow, with a brownish-red longitudinal stripe in the middle. Antennæ: basal joint reddish; third joint reddish-brown; arista black; palpi reddish-yellow, with short black pile. Thorax brownish-yellow, with black bristles and a shorter soft yellowish pubescence; on the dorsum, four black lines are perceptible; the intermediate pairs diverge posteriorly, and do not reach much beyond the suture; the lateral lines are broadly interrupted at the suture, and do not reach either the anterior or the posterior margin. Scutellum nearly of the same color with the thorax, with numerous black spines. Abdomen bright ferruginous red, with black spines, and a shorter, dense, rufous pubescence, especially perceptible posteriorly; on the first segment, under the scutellum, there is a triangular black spot; in some specimens, this spot encroaches slightly upon the second segment; sometimes there are similar triangular spots in the middle of the third and fourth segments, the spot on the fourth segment being occasionally very large; the spot on the third segment is entirely wanting in the majority of my specimens.. Legs yellowish-red, with black bristles and yellow pile on the femora. Wings tinged with brownish; the veins reddish-yellow near the base.

Hab.—Rocky Mountains in Colorado, common. I found it very commonly about Georgetown, Colo., at an altitude of 8–9,000 feet. Among the described *Dejeaniæ, D. rufipalpis* Macq. from Mexico seems nearest to it.

Family ORTALIDÆ.

PYRGOTA DEBILIS n. sp. ♀—Brownish; wings mottled with numerous pellucid spots; front yellow, tinged with brownish; ovipositor almost of equal breadth, blunt at tip. Length (including ovipositor) 7–8^{mm}; wing 8–9^{mm}.

Head pale yellow; front above the antennæ with a tinge of reddish-brown; sides of front and vertex yellow, the latter with a black dot in the middle; occiput yellow, with a brown spot in the shape of a W above the neck; antennal foveæ separated by a brown ridge, which is forked in front and connected with brown lines separating the sides of the face from the middle. Antennæ yellowish-brown; first joint brown; arista 2-jointed, but first joint extremely small. Thorax pale yellow; a double brown stripe in the middle, abbreviated behind; a lateral brown stripe on each side, abbreviated in front and interrupted at the suture; a humeral brown dot; a pleural, irregular brown stripe, running from the neck to the root of halteres; a pectoral brown spot below it. Scutellum yellow; metathorax brown, with a yellow line in the middle Abdomen brownish-yellow, more brown on the sides, narrow, almost. linear. Ovipositor about two-thirds of the abdomen in length, brownish-yellow, with brown margins; it is of nearly equal breadth, the tip being broad and bluntly truncate; on its anterior portion, on both sides, there are shallow depressions, with slightly projecting corners under them. Feet brownish-yellow; femora tipped with black; tibiæ with brown rings a little beyond the middle, which are pale on the four anterior and more distinct on the hind tibiæ. Wings pale brown, densely mottled with pale dots; a short brown band connects the small cross-vein with the costa; posterior cross-vein more .oblique than in *P. valida* Harris, and last section of fourth vein much less arcuated.

Hab.—Bee Springs, Kentucky (F. G. Sanborn). Two females.

This species is very like *P. valida* Harris in its general appearance, but much smaller; head and feet have a different coloring, the ovipositor a totally different structure, the posterior cross-vein a different position, etc. It cannot well be *Oxycephala maculipennis* Macquart, which is larger, etc.

Family TRYPETIDÆ.

TRYPETA (ŒDICARENA) PERSUASA n. sp., ♂.—Wings like Monographs, etc., iii, tab. xi, f. 15, except that the infuscated border of the apex is prolonged along the anterior margin, so as to come in contact with the cross-band at the end of the second longitudinal vein. The brown spot on the pointed end of the anal cell is much larger; the basal portion of the wing, including the costal cell, but excluding the hyaline inside of the second basal cell, is yellow. Head yellow; front bright gamboge-yellow, with a silky reflection. Antennæ reddish-yellow; arista yellowish at base; frontal bristles black. Thorax reddish-yellow, with a grayish pollen, somewhat concealing two indistinct longitudinal brown. ish stripes, expanded anteriorly, and bearing two darker spots posteriorly; a large black spot on each side between the root of the wings and the scutellum; two smaller black spots at the base of the scutellum. Halteres yellow. Abdomen ferruginous-red, with a slight grayish-yellow pollen clothed with recumbent black pile; no longer bristles. Legs reddish-yellow. Length about 6^{mm}.

Hab.—Denver, Colo. (P. R. Uhler and A. S. Packard, in July). Two male specimens. The description of the thorax was drawn from Mr. Uhler's specimen; on the other, the stripes and spots upon it were much less visible.

Observation.—This species is most closely allied to *T. tetanops* Loew

(Monogr., iii, 245) from Mexico, for which the subgenus *Œdicarena* was established. The peculiar structure of the large head, proboscis, eyes, the short wings, the straight course of the third vein, etc., are all to be found in *T. persuasa*, as described in *T. tetanops*. This is the second *Œdicarena* known, and the first in the United States, therefore an interesting addition to the fauna.

TRYPETA (EUARESTA ?) sp.—Very like *T. æqualis* Loew (Monogr., i, 86, and iii, 308, tab. x, f. 20), but probably different; on the front femora above, a black stripe, not mentioned in Dr. Loew's description, and not visible in my eastern specimens. The wings are broader and the hyaline spots on them larger. Cañon City, Colo. (P. R. Uhler).

TRYPETA (EUTRETA) SPARSA, Loew, Monogr., i, 78, iii, 274, tab-x, f. 13.—A specimen from Manitou, Colo. (Uhler), and another from Southern California agree in the main with the specimens from the Atlantic States. (Compare, however, what Dr. Loew says about this species in the 3d volume of the Monographs).

TRYPETA (*sensu strict.*) sp.—A single female from Colorado Springs (P. R. Uhler) is very like *T. palposa* Loew (Monogr., iii, 253, tab. x, f. 9). The picture of the wings is like the quoted figure, only the brown cross-bands covering the two cross-veins are *not* connected on the fifth vein in my specimen. As Dr. Loew's specimen was a male, and indifferently preserved, it will be more prudent to wait for more material.

TRYPETA (ASPILOTA) ALBA Loew, Monogr., iii, 285, tab. xi, f. 11.—Cañon City, Colo. (P. R. Uhler). I observe, however, that the third antennal joint is not *round*, as stated in the description, but has a distinctly marked angle at the end.

TRYPETA (ENSINA) HUMILIS Loew, Monogr., i, 81, iii, 291, tab. x, f. 17.—Cuba (Lw.); Key West; the Bermudas; Denver, Colo. (Uhler).

TRYPETA (STRAUSSIA) LONGIPENNIS Loew, Monogr., i, 65, iii, 238, tab. x, f. 2 ♂, 3 ♀.—Different localities in Southern Colorado in June (W. L. Carpenter). The singular variety *longitudinalis* Lw. also occurred there; also in Golden, Colo., July 3 (A. S. Packard).

TRYPETA (URELLIA) sp.—Very common at Crafton, near San Bernardino, in Southern California, in March. Resembles *T. solaris* Loew (Monogr., i, 84, iii, 325, tab. x, f. 19) very much. The spot on the wings of the female is almost exactly like the figure, which is also taken from a female; at the same time, that spot is subject to considerable variations in different specimens. In the male the spot is smaller, and the two rays running toward the apex, as well as that reaching toward the stigma, are not to be found; but in this sex likewise it is difficult to find two specimens absolutely alike. A small gray spot on the fifth vein, mentioned in the description of *T. actinobola* Lw., sometimes, but not always, makes its appearance here. A difference which seems to be constant lies in the fact that there is no brown around the small cross-vein, nor any dot on its proximal side nor in the discal cell. This species seems to have a wide distribution; I have a female from Santa Monica,

Cal., and another, exactly similar, from Colorado Springs (Uhler). Before describing this species, it will be necessary to compare it to the original specimens of *T. solaris* and *actinobola*. *Trypeta femoralis* Thomson (Eug. Resa, 582) is an *Urellia*, and may be only a variety of my species.

TRYPETA (TEPHRITIS) FINALIS Loew, Monogr., iii, 296, tab. xi, f. 4 (California and Texas).—Very common about Lake Tahoe and Webber Lake, Cal. (July 18–22), on *Wyettia mollis*, a Composite, which is evidently its food-plant.

TRYPETA (ACIDIA) FAUSTA n. sp., ♂ ♀.—Black; head, lateral stripes of thorax, scutellum, tibiæ, and tarsi yellowish or yellow; wings nearly like those of *A. fratria* (Monographs of North American Diptera, iii, tab. x, f. 4), but their proximal third hyaline with a narrow black crossband. Length 4–5ᵐᵐ.

Face, palpi, proboscis, and occipital orbit pale yellow; antennæ and lower portion of front orange-yellow; sides of front pale yellow, its upper portion ferruginous yellow, grayish-pollinose; occiput blackish. Thorax black; dorsum clothed with a short golden-yellow pubescence and a grayish pollen, forming two broad stripes; the longer bristles are black; a yellow stripe between the humerus and the root of the wing; scutellum pure yellow. Abdomen black, with black hairs. Legs: coxæ blackish, yellowish at the tip; trochanters clay-yellowish; femora black, their tip clay-yellowish; tibiæ and tarsi clay-yellowish. Halteres yellowish. The base of the wings is as in *A. suavis*, the apical portion like that of *A. fratria* (Monographs, etc., vol. iii, tab. x, fig. 10, and f. 4). The basal third of the wing is hyaline, tinged with yellow at the root and on the veins, and with a narrow black cross-band, beginning at the humeral cross-vein and ending on the sixth longitudinal vein (see fig. 10); the black color begins exactly where it does in fig. 10, and incloses a hyaline triangle reaching from the costa to the interval between the third and fourth veins; a hyaline spot is inclosed by the black on the distal part of the discal cell (as in fig. 4, only smaller); the black region emits a cross band toward the posterior border, parallel to the black border, running along the apical portion of the costa (as in fig. 4); the sinus between that cross-band and the black costal border is less deep than in fig. 4, and only reaches the third vein; sometimes it is surmounted by a small hyaline dot. The distance between the two cross-veins is as in fig. 10,—that is, the small cross vein is about the middle of the discal cell; the anal cell is not drawn out in a point at all.

Hab.—Mount Washington, alpine region (George Dimmock). A male and a female.

TRYPETA (ŒDASPIS) PENELOPE, ♂ ♀.—Reddish-yellow; upper side of thorax black, clothed with short, coarse, yellowish bristles; scutellum and metathorax black, shining; wings with three brownish cross-bands; cross-veins very approximate. Length about 4ᵐᵐ.

Head and antennæ reddish-yellow; cheeks pale yellow; thorax black

and shining above, clothed with short, coarse, yellowish or reddish (♀)
bristles; under side of the thorax pale yellowish; scutellum black, shining,
with four bristles; metathorax black, shining; abdomen reddish; ovipositor reddish, not longer than the last abdominal segment; feet reddish-
yellow. The cross-bands on the wings are somewhat like Monographs,
etc., vol. iii, tab. xi, fig. 17 or 18; first a short one, starting from the
humeral cross-vein; next two bands forming an inverted V, the angle
of which rests on the anterior margin, the two ends on the posterior one;
finally, a band along the apical margin, coalescent at one end with
the angle of the V, and ending on the other side in the second posterior
cell, close beyond the tip of the fourth vein. The color of these bands
is reddish-brown; the ends of the V and the posterior portion of the
apical band are darker brown. There is a small brownish cloud on the
posterior margin between the two branches of the V, and a very narrow hyaline space between the apical band and the costal vein near
the tip of second longitudinal vein; sometimes this space is reduced to
a small spot only (at any rate, this hyaline space is much smaller than
in the above-quoted figures).

Hab.—Manlius, N. Y. (J. H. Comstock).

Observation.—This species has all the characters of a true *Œdaspis;*
approximate cross-veins, the posterior one especially being very oblique;
a black, shining scutellum; a short ovipositor, etc.

TRYPETA (EUTRETA) DIANA n. sp., ♂ ♀.—Black, shining; abdomen red,
except the tip, which is black; wings broad, black, with whitish drops,
the apex margined with white. Length, ♂ 5ᵐᵐ; ♀ (with the ovipositor)
6–7ᵐᵐ. Length of wing, ♂ 3ᵐᵐ; ♀ 4ᵐᵐ.

Front broad, reddish, with black bristles on the sides; the usual
bristles on the vertex likewise black; but besides these there are numerous stubble-shaped whitish (♂) or whitish yellow (♀) bristles on
the vertex and on the posterior orbit above. Antennæ pale yellowish;
arista, except at the basis, black. Face whitish-pollinose. Thorax black,
shining, clothed above with short whitish hairs, not dense enough to
interfere with the luster of the dorsum. Legs black, shining; tarsi pale·
Abdomen blood-red. The two last segments in the male and the ovipositor in the female black. The penultimate segment in the male shows
a narrow, reddish, posterior border. The first joint of the ovipositor is
about equal to the two last segments taken together. The wings are
broad, rounded, black, covered with numerous white round dots, rather
uniformly spread over the disk of the wing, but not encroaching upon a
rather broad border, which is uniformly black. At and beyond the tip
of the sixth longitudinal vein, however, the white dots reach the margin
of the wing. The apex of the wing between the tip of the second vein
and the middle of the second posterior cell has a crescent-shaped white
border. The posterior cross-vein is very oblique and only gently curved.

A male and a female specimen were bred by Mr. C. V. Riley from a

11 H B

gall on the Wild Sage (*Artemisia tridentata*) in Missouri. The flies issued June 7 and 9.

Observation.—The wing of this species is a'together like that of *Trypeta sparsa* Wied. (Monographs, etc., vol. iv, tab. x, f. 13), only the coloring is darker brown; the white dots are less dense; there is no white mark at the tip of the first longitudinal vein; the white apical crescent begins exactly at the tip of the second vein, having only a yellowish prolongation before it. In the male specimen, the white drops are very faint, and disappear entirely in the surroundings of the fifth longitudinal vein.

TRYPETA (ZONOSEMA) BASIOLUM n. sp., ♂ ♀.—Yellow; somewhat ferruginous on the thorax; metathorax with two black spots; wings not unlike Monographs, etc., iii, tab. xi, f. 15, only the brown border on the apex is prolonged anteriorly, so as to come in contact with the cross-band. Length 5–6^{mm}.

Yellow; somewhat ferruginous on the thorax and front; antennæ yellow, reaching beyond the middle of the face; third joint elongated, its upper edge straight, perhaps even slightly concave; its tip angular; arista brown, yellowish at base, finely pubescent; bristles on the head black; two black spots on the metathorax under the scutellum; they are rounded in the male, larger and in the shape of a longitudinal stripe in the female; a small black dot immediately behind the root of the wing, I perceive only in the female. Abdomen brownish-yellow, with black pile; ovipositor broad, inverted-trapezoidal, shorter than the two last segments taken together, yellowish-ferruginous. Wings subhyaline; a short, oblique, pale brown mark in the costal cell near the humeral cross-vein; a pale brown cloud in the innermost proximal end of the first basal cell; a similar pale cloud on the angular vein closing the anal cell; a brown cross-band runs from the anterior margin, covers the small cross-vein, and stops short in the middle of the third posterior cell without reaching the hind margin; the anterior end of this cross-band is very dark brown, and nearly fills out the interval between the ends of the auxiliary and first veins; a second cross-band begins at the distal end of the marginal cell, which it fills out, except its extreme tip, covers the great cross-vein, and ends, on the posterior margin of the wing, in the distal end of the third posterior cell; in the middle, between these two principal cross-bands, on the anterior margin, there is a short, oblique, brown streak, which bisects the marginal cell, and is prolonged as a pale shadow across the submarginal; along the apex of the wing, there is a brown border, which begins at the second cross-band and ends a little beyond the fourth vein.

Hab.—Brookline, Mass. (Fred. C. Bowditch). Two specimens.

Observation.—Meigen's tab. 43, f. 16, very nearly represents the picture of the wings of this species, only the cross-bands in my species are farther apart toward their end. Meigen's figure represents the wing of *Zonosema alternata* Fall. (syn. *continua* Meig.), and I believe that *T. basiolum* must likewise be placed in the subgenus *Zonosema*.

GENERAL REMARKS ON THE DIPTERA OF THE WESTERN REGION, AND OF CALIFORNIA IN PARTICULAR.

In the introductory paragraphs to the families of *Diptera*, of which I have treated in the preceding pages, I have attempted some generalizations concerning the character of the Western, and especially of the Californian, faunæ, and their relationship to other faunæ. In trying now to sum up these generalities, I become more than ever aware of the insufficiency of our present knowledge of the *Diptera* of that fauna, and of the meagerness of the results obtained. If I persist, nevertheless, in my attempt, it is because I find that the general results thus far reached for the order of *Diptera* coincide with those obtained in the other orders of insects, and that their publication, even in their present imperfect form, may tend to confirm the accuracy of those results.

The belief of many, and under which I confess to have labored until better informed, that the Rocky Mountains form a natural boundary for a distinct entomological fauna, is erroneous. It is a well-known fact that somewhere between the Rocky Mountains and the Mississippi there is a line, west of which agriculture becomes precarious without artificial irrigation. This line, which some observers place about longitude 98°, marks the eastern limit of a region which extends to the Pacific Ocean, and is characterized by peculiar conditions of life and a peculiar fauna. Among these conditions, the principal, the one which determines the most striking features of the whole region, is *summer dryness*. The natural limits of this region, both north and south, are countries where summer rains prevail. In the north, this limit marks the beginning of northern forms, some of which are circumpolar; in the south, the advent of a tropical fauna. All living beings, and the insects among the rest, have to adapt themselves to that condition of dryness. This explains the prevalence of *Heteromera* among the Beetles the remarkably stout carapace of which enables them to withstand desiccation for a surprisingly long time (in Lacordaire's collection, an *Eleodes* remained alive on its pin for seven months, of course without any food). Such *Heteromera* escape the heat of the day by their nocturnal habits. Their usually black color is the concomitant of such habits. Certain *Carabidæ*, also nocturnal, have the same black color, and often a remarkable resemblance to the *Heteromera* in their outward appearance. Dry soil and sunny exposures attract the burrowing *Hymenoptera*—Bees, Sand-wasps, *Mutillæ*—which form another characteristic feature of the region. The nests of these are infested by numerous parasites—the *Meloidæ* among *Coleoptera*, the *Bombylidæ* among *Diptera*. Such is the explanation of the presence in the Western Plains of numerous species of *Cantharis* (*Epicauta*) and of the *Bombylidæ*, which prevail among the *Diptera* of the region as much as the *Heteromera* among the Beetles.

The same conditions of life, with the same results, exist in other continents. There is a vast region in the Old World which resembles, in that respect, the North American western region. It is the so-called

Mediterranean and Central Asiatic region, extending from Portugal to Turkestan, and embracing Southern Europe and Northern Africa. It is also characterized by the prevalence of *Heteromera* among the *Coleoptera* and of *Bombylidæ* among the *Diptera*. It displays a remarkable unity of character through that vast expanse of country. The very striking genus *Julodis* (*Buprestidæ*) occurs in Spain as well as in Turkestan, and nowhere else, except at the Cape of Good Hope. The small family of *Glaphyridæ* (*Lamellicornia*) is almost exclusively confined to the same Mediterranean region, and also, although in other forms to the Cape. The genera *Cleonus* and *Brachycerus* (*Rhyncophora*) living on sandy soil and in hot situations characterize the same Mediterranean and Central Asiatic region.

The same unity of character distinguishes the North American western fauna. Besides the *Melasomata* and other *Heteromera*, which occur in increasing numbers from the Plains of Kansas and Colorado to California, the entomologist is struck by the occurrence of other forms of *Coleoptera*, unknown in the Atlantic States; for instance, the *Dasytidæ*, which occur in Colorado and in California, and are also represented in Europe. *Masaris*, a very peculiar genus of *Vespidæ*, for a long time known only from Algiers, has been found since in the Rocky Mountains, in Texas, and in California. A number of *Odonata* occur in California, in the Yellowstone region, and in Colorado, but do not extend farther east. Among the *Diptera*, I will quote some leading species, as *Tabanus punctifer*, *Silvius gigantulus*, *Eupeodes volucris*, *Lordotus gibbus*, the genus *Ospriocerus*, which occur in the whole western region, and not in the Atlantic States; *Lordotus*, *Eupeodes*, and *Ospriocerus*, being new genera, as far as known, peculiar to that region; *Silvius*, a European genus, which, if it occurs at all in the Atlantic States, must be exceedingly rare. The very remarkable case of *Syrphus pyrastri*, a European Syrphid, very common in California, and also found in Colorado and New Mexico, but never east of the Mississippi, will be discussed below.

The resemblance between the western and the Mediterranean and Central Asiatic fauna, is an analogy, due to the identity of meteoric conditions; it is not a relationship. The same families of insects will prevail, not necessarily the same genera. Thus, among the *Heteromera*, the great majority of genera in both regions are different. The *Diptera* are more cosmopolitan in the distribution of their genera. Such large genera as *Bombylius*, *Anthrax*, *Stenopogon*, *Saropogon*, belonging to dry regions, are the same in the Old World and in North America; but, in the same families, *Bombylidæ* and *Asilidæ*, a number of small genera occur, peculiar to each region. Cases of identity of small and exclusive genera, like the above-quoted one of the Vespid *Masaris*, are for this reason very interesting. *Pedinocoris brachonyx* Mayr is a large aquatic *Hemipteron*, of which I brought specimens from San Diego, Cal. Mr. Uhler tells me that the same genus is known to occur in Egypt and Turkey. Among the *Diptera*, I will name the small genus *Xestomyza*, a singular Therevid represented in the Mediterranean fauna, at the Cape, and in California.

Altogether different from that analogy, arising from the similarity of meteoric conditions, are certain resemblances between the western fauna (and especially that of California) and the fauna of Northern and Central Europe, resemblances manifested in cases of generic and even specific identity. These cases derive their significance from the fact that they are foreign to the fauna of the Eastern United States; and they are the more strange, as, far from being favored by any similitude of meteoric and botanic conditions, they seem to exist in spite of differences in these conditions. Northern and Central Europe, in their climate and the character of their vegetation, are certainly more like the northern Atlantic States of the Union than the western region. In such cases, faunal resemblances are more than analogies, and seem to indicate some relationship, some hidden genetic connection between the faunas of Europe and the western portion of this continent.

A European who has lived for some time in the Eastern States of the Union, and crosses the Rocky Mountains for the first time, is soon struck by the appearance of the Magpie, a European bird unknown in the Eastern States. I am told that many parallel cases occur among birds. Similar coincidences occur in all the orders of insects.

Plusia gamma is very common in California, and also occurs in Colorado and Texas; not in the Atlantic States. It is a well-known European species. The genus *Parnassius* occurs in the Coast Range, the Sierra Nevada, and the Rocky Mountains; it is found in the Alps in Europe, in Sweden, and in Finland; it is not found east of the Mississippi. *Argynnis*, *Melitæa*, *Lycæna*, and the *Satyridæ*, are by far more common in California and in Europe than in the Eastern States. The Californian *Papilio zolicaon* is almost the same as the European *P. machaon*. In Mr. Grote's Check List, etc. (p. 22), I find the statement that *Ochria saucelitæ* Grote is a *Noctua*, with a horned clypeus, like the European *Ochria flavago* Hübner, and that no similar case is known to occur in the Eastern States.

Among the *Neuroptera*, there is the European genus *Rhaphidia*, quite common in California, and occurring in several species; it is unknown in the Atlantic States. Among the *Orthoptera*, the genus *Locusta* occurs in Europe and in the Western Region, and not in the Atlantic States. Among the *Diptera*, I found in the Yosemite Valley a species of the genus *Elliptera* (*Tipulidæ*), a genus discovered in Europe within the last fifteen years only, and not known to occur in the Atlantic States. A species of the European genus *Silvius* is common in California, and also occurs in Colorado; I have never seen a *Silvius* taken east of the Mississippi, although one is described by Wiedemann. The genus *Sphærophoria* (*Melithreptus* Loew, *Syrphidæ*) is more abundantly represented in California and in Europe than in the Atlantic States. The Californian *Leptidæ* have a more European general appearance than those of the Atlantic States. The above-mentioned *Syrphus pyrastri* is a common European insect, the larvæ of which live on *Aphides;* it is quite common in all parts of California; I also have specimens from Utah, Colorado, and Northern New Mexico. To my knowledge, it has never been found east of the Mississippi. The suggestion that *S. pyrastri* may have been accidentally introduced in California, and is gradually spreading eastward, may be met by the fact that Say's *Syrphus affinis*, which is nothing else but *S. pyrastri*, was caught by that entomologist near the Arkansas River as early as 1820, and does not seem to have advanced eastward since. The occurrence of this species in the west gains a peculiar significance from its simultaneous occurrence in Chili, recorded by Macquart.

Not all the coincidences with the European fauna just alluded to belong to the whole western fauna. Many are peculiar to California only, although, owing to our imperfect knowledge of the western *Diptera*, we are often unable to state which among them belong to the one or to the other category.

The affinities with the Chilian fauna seem to be especially Californian. Besides the case of *Syrphus pyrastri*, just mentioned, the following instances have occurred to me:—The Tipulid *Protoplasta vipio*, from California, belongs to a remarkable group, hitherto represented by three species only: *Macrochile spectrum*, a fossil *Dipteron* from the Prussian amber; *Protoplasta fitchii*, from the Atlantic States; and *Tanyderus pictus*, from Chili. A somewhat analogous case is that of *Eriocera californica* (*Tipulidæ*), one of the *Erioceræ*, with enormously prolonged antennæ in the male. Of such *Erioceræ* I have hitherto known only three species from the northern United States, two fossil species in amber, and one from Chili (the *Megistocera chilensis* of Philippi, which I strongly suspect to be an *Eriocera*). My new genus *Rhaphiomidas* (*Midaidæ*) has its nearest relative in *Mitrodetus* from Chili. The genus *Clavator* (*Asilidæ*) from Chili, if my identification be correct, is represented in California. The most interesting case is that of *Apiocera*, an anomalous genus, intermediate between *Asilidæ* and *Midaidæ*, and hitherto found only in Chili and Australia. I describe a species from California.

Several genera of *Diptera* have not been yet found outside of the limits of California, although it is very probable that they have a somewhat wider distribution. Such are *Eulonchus* (*Cyrtidæ*), *Dicolonus*, *Ablautatus* (*Asilidæ*), *Pantarbes*, *Paracosmus* (*Bombylidæ*), the extraordinary genus *Polymedon* (*Dolichopodidæ*), and *Phyllolabis* (*Tipulidæ*).

Ospriocerus (*Asilidæ*), *Lordotus* (*Bombyl.*), and *Eupeodes* (*Syrphidæ*) have already been named as peculiar to the whole western region.

Among the singularities of the Californian fauna of *Diptera* I will mention the apparent rarity of *Trichocera* (*Tipul.*), of which I found only a single specimen of a rather peculiar species; the apparently frequent occurrence of *Hygroceleuthus* (*Dolichop.*), of which I found two species, before I had collected more than one *Dolichopus* (in the Eastern States, a single species of *Hygroceleuthus* is known, and some fifty species of *Dolichopus*); the large number of *Tipulæ* and the comparatively rare *Pachyrrhinæ*;

the occurrence of *Trimicra pilipes*, apparently identical with the European and probably with the North American *T. anomala*, although the latter is comparatively rare in the Eastern States, while *T. pilipes* is exceedingly common in all California in winter.

In the whole western region, the genera *Tabanus* and *Chrysops* seem to be far less abundant in species than in the region east of the Mississippi.

Of the anomalous family *Blepharoceridæ*, all the species of which seem to be rare and local, I have described a species from Yosemite Valley and a new genus from the Rocky Mountains.

After having detailed the peculiarities of the western, and especially of the Californian, Dipterous fauna, it remains for us to examine what they have in common with the eastern fauna. As a rule, cases of specific identity between those regions occur more frequently in those same families in which cases of specific identity are more frequent between Europe and North America. Several Californian *Limnobiæ* are not distinguishable from eastern species. *Trimicra pilipes*, already mentioned, and *Symplecta punctipennis*, seem to be species of nearly universal occurrence. Several *Syrphidæ*, common in the Eastern States, also occur in California. *Asilidæ* and *Tabanidæ*, on the contrary, seem to be different in both regions, just as no species of these two families is as yet known to be common to North America and Europe.

The genera *Ceraturgus*, *Nicocles* (*Asilidæ*), *Triptotricha* (*Leptidæ*), and the singular *Epibates* (*Bombylidæ*), are worth noticing as being common to both sides of North America, and not found yet outside of that continent. The remarkable genus *Rachicerus* (*Xylophagidæ*) belonged in the same category, until recently, when it was found in Spain.

In the mountain-ranges which cross the western region from north to south, some northern and subarctic genera and species are able to reach very far south, and thus to come in contact with the forms of the local fauna. In Yosemite Valley, at an altitude of 4,000 feet, the mixture of truly Californian forms with those peculiar to the Sierra is only beginning, the latter being comparatively rare. Around Webber Lake, that is, farther north, and at an altitude of 7,000 to 8,000 feet, Californian genera and species still occur in abundance, but more northern forms are frequently met with them. The northern genus *Scellus* (*Dolichopodidæ*) occurs alongside of the Californian *Eulonchus* (*Cyrtidæ*). With the Californian *Dasyllis astur* (*Asilidæ*) and *Laphria vultur* (*id.*), I found *Laphria rapax* (*id.*), which looks like a northern form, although I may be mistaken in my surmise. The specimens of *Dasyllis astur*, found at that altitude, have much more yellow pile on their legs, neck, and pleuræ than those which were taken but little above sea-level. According to the same law, *Dasyllis flavicollis* Say, which ranges from Canada to Texas, has much more yellow on its legs and pleuræ in the north than in the south. Many interesting species were found round Webber Lake:

I will name a new *Tachytrechus* (*Dolichop.*), related to *T. mœchus* of the Eastern States, which I used to find abundantly near the Trenton Falls, New York; a new *Sphecomyia* (*Syrphidæ*), a remarkable genus, of which only two species were hitherto known, one in Europe and the other in North America, and those two may yet turn out to be identical; thirteen species of the genus *Cyrtopogon* (*Asilidæ*), eleven of which were undescribed, and some of them remarkably handsome (in Dr. Schiner's Catalogue of *Asilidæ*, published in 1866, only thirteen species of *Cyrtopogon* are enumerated for the whole world). The other orders of insects afforded the same interest. *Parnassius* was very common; two new species of *Cicada* were found, etc.

Of the fauna of the Rocky Mountains, I had occasion to speak in another place (Report on the Diptera collected by Lieutenant Carpenter in Colorado in 1873, in the Annual Report of the United States Geological and Geographical Survey of the Territories for that year). The relationship of the fauna in the higher regions of those mountains to that of the northern latitudes of the continent is much more marked than that of the fauna round Webber Lake in the Sierra. A series of characteristic northern forms were found in Lieutenant Carpenter's collection:—*Hesperinus brevifrons* (*Bibionidæ*), which had been received from Mackenzie River and collected by myself on Mount Washington; *Arctophila flagrans, Tipula macrolabis, Helophilus bilineatus,* etc. For want of time, I did not collect much in the Rocky Mountains, but was struck by the frequent occurrence, near Georgetown, Colo. (8,500 feet altitude), of a species of *Dejeania* (*Tachinidæ*), a genus which was hitherto received from South America and Mexico. Near Manitou, Colo. (altitude 6,400 feet), another very large and peculiar Tachinid occurred, of which I also have specimens, collected by Mr. Cleveland near San Diego.

Such facts, as well as many others mentioned in the course of the present paper, prove that there is a great deal to be learned yet about the laws regulating the geographical distribution of insects. In the mean while, it is useful to keep such facts in view by singling them out from the arid mass of descriptive entomology.